Control of leaf growth

Edited by

N. R. BAKER

Department of Biology, University of Essex

W. J. DAVIES

Department of Biological Sciences, University of Lancaster

C. K. ONG

Department of Physiology and Environmental Science,
School of Agriculture, University of Nottingham

The right of the University of Cambridge to print and sell all manner of books was granted by Henry VIII in 1534. The University has printed and published continuously since 1584.

CAMBRIDGE UNIVERSITY PRESS

Cambridge

London New York New Rochelle

Melbourne Sydney

Published by the Press Syndicate of the University of Cambridge
The Pitt Building, Trumpington Street, Cambridge CB2 1RP
32 East 57th Street, New York, NY 10022, USA
10 Stamford Road, Oakleigh, Melbourne 3166, Australia

First published 1985

Printed in Great Britain at the University Press, Cambridge

Library of Congress catalogue card number:

British Library cataloguing in publication data

Control of leaf growth.
1. Leaves—Growth
I. Baker, N.R., II. Davies, W.J. III. Ong, C.K.
581.3′1 QK649

ISBN 0 521 30480 6

CONTENTS

Contents

LIST OF CONTRIBUTORS

Dr C.K. BAKER
Department of Physiology & Environmental Sciences,
University of Nottingham, Sutton Bonnington,
Leicester LE12 5RD

Dr N.R. BAKER
Department of Biology, University of Essex,
Colchester CO4 3SQ

Professor R.E. CLELAND
Department of Botany, University of Washington,
Seattle, Washington 98195, U.S.A.

Dr J.E. DALE
Department of Botany, University of Edinburgh,
Edinburgh EH9 3JH

Dr W.J. DAVIES
Department of Biological Sciences,
University of Lancaster, Bailrigg, Lancs. LA1 4YQ

Dr J. DIGBY
Department of Biology, University of York,
York YO1 5DD

Dr R.D. FIRN
Department of Biology, University of York,
York YO1 5DD

Dr J.N. GALLAGHER
Lincoln College, Canterbury, New Zealand

Professor T.C. HSIAO
Department of Land, Air and Water Resources,
University of California, Davis,
California 95616, U.S.A.

Dr J. JING
Department of Land, Air and Water Resources,
University of California, Davis,
California 95616, U.S.A.

Dr H.G. JONES
East Malling Research Station, Maidstone,
Kent ME19 6BJ

Dr D.W. LAWLOR
Department of Physiology & Environmental Physics,
Rothamsted Experimental Station, Harpenden,
Hertfordshire AL5 2JQ

Dr J.E. LEACH
Department of Physiology & Environmental Physics,
Rothamsted Experimental Station, Harpenden,
Hertfordshire AL5 2JQ

Professor R.M. LEECH
Department of Biology, University of York,
York YO1 5DD

Professor H.K. LICHTENTHALER
Botanical Institute, University of Karlsruhe,
Kaiserstrasse 12, D-7500 Karlsruhe, FRG

Professor T.A. MANSFIELD
Department of Biological Sciences,
University of Lancaster, Bailrigg, Lancs. LA1 4YQ

Dr C.K. ONG
Department of Physiology & Environmental Sciences,
University of Nottingham, Sutton Bonnington,
Leicester LE12 5RD

Dr M.G. SCHMIDT
Department of Botany, University of Washington,
Seattle, Washington 98195, U.S.A.

Professor W.K. SILK
Department of Land, Air and Water Resources,
University of California, Davis,
California 95616, U.S.A.

Professor E. STEUDLE
Arbeitsgruppe Membranforschung am Institut fur
Medizin der Kernforschungsanlage Julich, Postfach
1913, D-5170 Julich, FRG

Dr A.D. TOMOS
Adran Biocemeg a Gwyddor Pridd, Coleg y Brifysgol,
Bangor, Gwynedd LL57 2UW

Dr A. TREWAVAS
Department of Botany, University of Edinburgh,
Edinburgh, EH9 3JH

Dr E. VAN VOLKENBURGH
Department of Botany, University of Washington,
Seattle, Washington 98195, U.S.A.

Dr M.E. WHITMORE
Department of Biological Sciences,
University of Lancaster, Bailrigg, Lancs. LA1 4YQ

PREFACE

The growth of leaves has been the subject of extensive quantitative studies by plant biologists for many years, however the current understanding of how leaf growth is controlled is poor. It was the aim of the Environmental Physiology Section of the Society for Experimental Biology to address and discuss this important topic in a seminar held at the University of Wales at Cardiff from the 3rd to the 5th April, 1984. The invited speakers, who came from many different disciplines ranging from biochemistry to agronomy but who all had a common interest in leaf growth, were asked to review the current knowledge of the factors regulating leaf growth and raise what they considered to be important problems still to be solved in the quest for a better understanding of the control of leaf growth. This brief was extended to the speakers' written contributions that comprise this volume, with the proviso that each chapter should be written at a level suitable for advanced undergraduate and first-year postgraduate students. The contributions fall naturally into two sections; firstly a consideration of the intrinsic factors within a leaf that determine leaf growth and secondly an evaluation of the environmental factors that can modify leaf growth.

It is hoped that publication of this volume will be particularly timely because of considerable advances made in several fields of study relating to the growth of leaves. Comprehensive investigations of the synthesis of cellular components in developing leaves have recently been undertaken, while the contribution made by developing energy transducing systems to leaf growth has also begun to be elucidated. New techniques to investigate the biophysical properties of plant cell walls have been developed and there is now considerable interest in extending the use of this new technology. The water relations of growing cells and water transport through growing tissue are currently particularly controversial areas. A consideration of the effects of air pollution on the growth of leaves is also appropriate at this time. It is hoped that the book will be of use not only to students of plant and crop physiology but will also provide stimulation, because of its multidisciplinary approach, for research scientists investigating the control of leaf growth.

We take this opportunity to thank all the authors for their enthusiasm and, in the majority of cases, the prompt submission of their manuscripts in a form which minimised the task of the editors. We are also indebted to Mrs S. Stern of Computer Services Partnership for efficient production of the final typescript, which involved the insertion after typing of a very large number of symbols, superscripts and subscripts.

While the S.E.B. carried most of the organisational load for this meeting, both the Association of Applied Biologists and the British Ecological Society provided substantial financial contributions to bring speakers from overseas. It is a pleasure to acknowledge this assistance which enabled us to organise a truly international meeting which proved to be of interest to members of all three Societies. We would also like to thank Dr Lynton Incoll, the Convenor of the Environmental Physiology Group of the S.E.B., for his extensive help and relentless encouragement during the organisation and running of the seminar at Cardiff and throughout the preparation of this volume.

N.R. BAKER
W.J. DAVIES
C.K. ONG

THE PHYSICAL LIMITATIONS OF LEAF CELL EXPANSION

A. Deri Tomos
Adran Biocemeg a Gwyddor Pridd, Coleg y Brifysgol, Bangor, Gwynedd, Wales.

INTRODUCTION

Although cell division lays down the foundations of leaf architecture, the thousands of different but highly specific leaf shapes, each with its different cell types, air spaces and recognisable external form have their bases in the magnitude and direction of cell wall expansion at the level of the individual cell. An understanding of these processes is thus central to the study of plant development physiology.

Biochemical processes combine to form a structure that can harness such physical phenomena as osmosis, hydrostatic pressure and the viscoelastic properties of wall polymers to drive and regulate growth. In this review I shall attempt to describe cell growth in terms of these physical processes and to give an account of our current understanding of how the physical and mechanical properties of cell walls and membranes allow growth to occur.

The amount and timing of area and volume expansion of leaf cells has been described in detail and in context for *Xanthium* by Maksymowych (1973) where a meristematic cell of some 0.5 pl expands to a 13.4 pl epidermal cell or a 6.7 pl mesophyll cell respectively. These mature cells, however, are quite small in comparison with those of *Tradescantia* for example, where normal epidermal cells reach several nanolitres in volume, while such specialised cells as the leaf bladder cells of *Mesembryanthemum* reach volumes greater than a microlitre. (Giant algal cells - *e.g.* *Nitella, Valonia, Chara* and *Halicystis* - reach sizes several orders of magnitude larger still.)

Most of our information on the physical characteristics of growth has been gathered from tissues other than that of leaves. It is with caution that these studies may be extrapolated to a description of leaf cell growth, although at present our knowledge is sufficiently rudimentary for differences in detail to be probably insignificant. The extensive review literature on growth (Cleland, 1971, 1977, 1981; Hsiao, 1973; Hsiao et al, 1976; Jones, 1982; Masuda, 1978; Penny & Penny, 1978; Preston, 1974) emphasises the special interest in the mode of action of plant growth substances. The role of auxin has been most widely studied although the influence on growth of gibberellins and other growth substances is becoming increasingly understood. Of specific relevance to cellular growth, in this series Sellen (1980) has reviewed the mechanical properties of cell walls, while the hydraulic conductances of tissue was the subject of Boyer (1977).

AN EMPIRICAL MODEL OF CELL GROWTH

The simplest model visualises the cell as a solution surrounded by a semipermeable membrane and an elastic cell wall that is also capable of irreversible expansion. During growth three major processes must be considered.

1. The increase of cell surface area. This is the deformation of the existing wall together with an increase in the bulk of the wall due to the intercalation of new material all over its surface (Green, Erickson & Richmond, 1970; Preston, 1974; Roland & Vian, 1979). In some cases, such as root hair growth, only limited areas of cell wall extend (Green, 1969). Presumably the processes at these growing zones are no different from those of "typical" growing walls. A feature related to cell wall growth - the potential limitation of growth by plasmalemma and tonoplast extension - has recently been investigated for maize and suggested to be non-limiting (Quaite et al., 1983).

2. The increase in cell volume. Since mature plant cells are 85-90% water, this increase must be largely due to

water uptake into the cell, especially into the vacuole.

3. The maintenance of osmotic relations. Without this, as the volume expands the osmotic contents would become diluted and turgor pressure diminish. This would influence not only continued growth, but also the structural rigidity of the tissue (i.e. growing leaves would automatically wilt).

The widely accepted cycle of physical events that accompany cell growth can be traced back to the work of Julius von Sachs (1873) and Wilhelm Pfeffer (1881) and is summarised in Fig. 1. Controversy remains as to the order of some of the events (e.g. do some growth substances promote growth by decreasing cellular osmotic potential?) although

Fig. 1. Physical events that accompany irreversible cell growth.

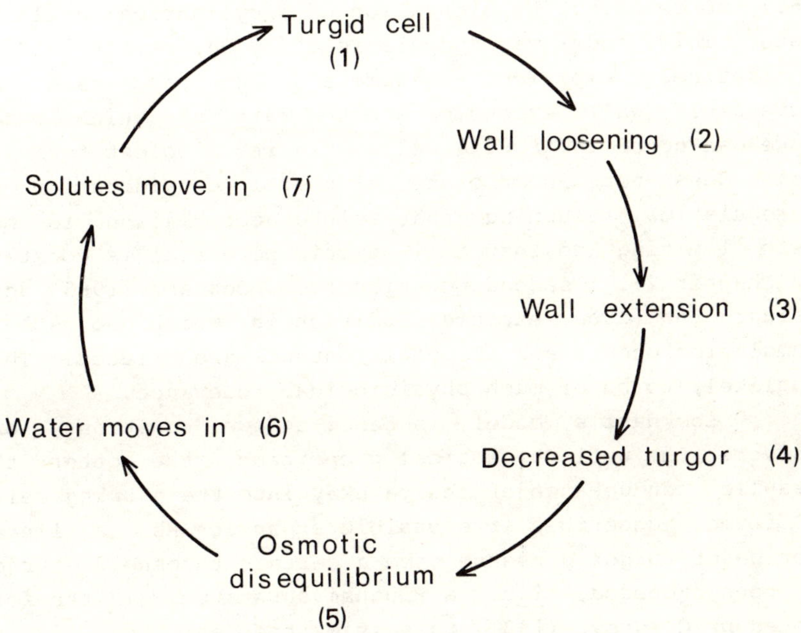

steps 3 to 5 are essentially simultaneous. These processes are represented here as a cycle, during *in vivo* growth all are presumably continuous. They are not independent, as I shall discuss below.

To describe these processes quantitively requires the definition of measurable physical parameters. To this end several comprehensive growth models have been formulated, *e.g.* notably one for maize root growth which incorporates some 27 independent variables (Grenetz & List, 1973). Other models include a membrane model (Hettiaratchi & O'Callaghan, 1974) and a fluid dynamic model (Silk & Erickson, 1979). As our understanding of the parameters extends such models will be used to advantage but our current state of knowledge of the individual parameters would appear to be insufficient to justify their use beyond theoretical discussion.

The most widely used model was proposed by Lockhart (1965) to describe the elongation of a cylindrical cell or tissue. This model was "popularised" by Ray *et al*. (1972) and defines empirical parameters for the rate of irreversible wall extension and the rate at which water uptake can occur in growing cells. In its simplest form the model does not incorporate a useful consideration of osmoregulation, assuming that solute accumulation is not growth limiting and that cell osmotic potential is constant (see Chapter 2). A second equation from Lockhart (1965) does consider the other extreme condition in which no solute accumulation occurs and the cell contents are diluted. This is unlikely to be of much physiological relevance.

Lockhart's model is based on two equations, one considers the wall rheological properties, the other the hydraulic conductance of the pathway into the growing cell. The former describes irreversible expansion as a linear function of turgor pressure once a certain threshold turgor has been exceeded, *i.e*. a Bingham substance. In the form adopted by Cosgrove (1981) this is written as:

$$\frac{1}{V_0} \cdot \frac{dV_0}{dt} = \phi\,(P - Y) \qquad (1)$$

(A key to all the parameters used in this review together with dimensions and synonyms is given in Table 1.)

V_0 rather than the more general V is used in equation (1) to avoid the effects of reversible (elastic) extension on the growth equation, although Lockhart (1965) shows that $dV_0/V_0 = dV/V$.

That extension can be a linear function of turgor after exceeding a yield threshold is supported by experimental observation from _Nitella_ (Green _et al._, 1971), _Avena_ coleoptile (Cleland, 1959), _Pisum_ roots (Gracean & Oh, 1972), radish cotyledons (Kirkham _et al._, 1972), soybean leaf (Bunce, 1977) and _Pennisetum_ leaf (Squire _et al._, 1983). It must, however, be emphasised that equation (1) is an empirical relationship. Nutting (1921), while describing the laws of deformation, observed that "experimentalists have rested in the belief that theoretical work had established the foundations of these laws, while theoretical writers, on the evidence of the experimentalists, have contented themselves with the first term of a Taylor series expansion of an unknown function". This still appears to be the case.

The second equation relates volume expansion to the hydraulic conductivity of the water pathway into the cell and incorporates a consideration of the laws of non-equilibrium thermodynamics (Katchalsky & Curran, 1965):

$$\frac{1}{V} \cdot \frac{dV}{dt} = Lp \cdot \frac{A}{V} \cdot (\sigma\Delta\pi - P) \qquad (2)$$

Although the theoretical basis of this equation is more sound than that of equation (1) it presents far greater experimental difficulties than the former, especially with regard to the surface area parameter and the homogeneity of Lp. For isolated cells in a well stirred bathing medium Lp, A and σ may be defined as the relevant homogeneous cellular

Table 1. The key to the abbreviations of the parameters used in the equations included in this review together with their dimensions and synonyms used by other authors. T = time, L = length, F = force, P = pressure (strictly $P = F.L^{-2}$), dml = dimensionless.

Abb.	Parameter	Dimensions	Synonyms
V	Cell volume	L^3	
V_o	V at zero turgor	L^3	
t	Time	T	
ϕ	Irreversible volumetric wall extensibility	$P^{-1}.T^{-1}$	m, WEx.V, E_g
P	Turgor pressure	P	P_{cell}, ψ_P, T
P_{wall}	Wall hydrostatic pressure relative to atmospheric	P	
Y	Yield stress threshold	P	ψ_P, th, P_c
A	Cell surface area	L^2	
Lp	Hydraulic conductivity	$L.T^{-1}.P^{-1}$	K, K_W
σ	Reflection coefficient	dml	
$\Delta\pi$	Osmotic potential step across membrane or tissue	P	
L	Volumetric hydraulic conductance (=(Lp.A)/V)	$P^{-1}.T^{-1}$	C, (L'p/V)
S	Stress	$F.L^{-2}$	σ
T_m	Maximum stress relaxation time	T	
T_o	Minimum stress relaxation time	T	
C	Stress remaining after T_m	$F.L^{-2}$	
b	(constant)	dml	
π_i	Intracellular osmotic potential	P	π_{cell}, ψ_s
S_T	Tangenial stress	$F.L^{-2}$	σ_T
S_L	Longitudinal stress	$F.L^{-2}$	σ_L
d	Wall thickness	L	t
r	Cell radius	L	

paramters and $\Delta\pi$ and P as the properties of vacuolar sap and bathing medium (Zimmermann & Steudle, 1978). For cells in a tissue, however, the situation may be very different, with hydraulic pathways with different properties occuring in parallel and in series (Boyer, 1977; Molz & Ferrier, 1982). The significance of σ to this problem is considered elsewhere in this volume by Steudle (Chapter 2).

The expression ($\sigma\Delta\pi$ - P) has often been equated with $d\psi$ (the step in water potential from one compartment to another). This is only strictly true if $\sigma = 1$, i.e. if the cells are behaving as ideal osmometers. This may not always be the case (Dainty & Ginzburg, 1964; Tyerman & Steudle, 1982).

In transpiring plants a further component must be taken into account. Transpiration will create a hydrostatic tension in the cell walls. This factor requires modification of equation (2) by replacing P with ($P - P_{wall}$), where P_{wall} represents the hydraulic tension of the wall. Thus, equation (2) may be abbreviated to

$$\frac{1}{V} \cdot \frac{dV}{dt} = L \cdot (\sigma\Delta\pi - [P - P_{wall}]) \qquad (3)$$

where L represents a "volumetric" hydraulic conductance for the cell and thus avoids some of the experimental difficulties innate in equation (2). Equations (1) and (3) may be equated for P to generate a general growth equation -

$$\frac{1}{V} \cdot \frac{dV}{dt} = \frac{\phi \cdot L}{\phi + L} \cdot (\sigma\Delta\pi - Y + P_{wall}) \qquad (4)$$

This is the partially empirical equation that may be a starting point for the definition of the physical parameters that govern cell growth. A feature of the equation is that although turgor pressure is central to growth it does not appear explicitly in the equation. Cosgrove (1983) has recently discussed this apparent paradox by showing that the turgor pressure of growing cells is not an independent variable but is a function of the L/ϕ ratio. Inspection of equation (4) shows that ϕ and L are algebraically symmetrical. The consequences of varying L/ϕ for growing cells is shown in Fig. 2. If L is limiting (i.e. $<<\phi$) then the turgor pressure of the cells will approach the yield stress threshold, below which no growth will occur. If, on the other hand, L is not limiting (i.e. $>>\phi$) then the cell will be close to osmotic equilibrium and turgor pressure will approach ($\sigma\Delta\pi + P_{wall}$). If L is not limiting then equation (4) will reduce to equation (1) (Ray et al., 1972).

Fig. 2. Influence of varying ϕ/L ratio on the turgor pressure of growing cells (after Cosgrove, 1983).

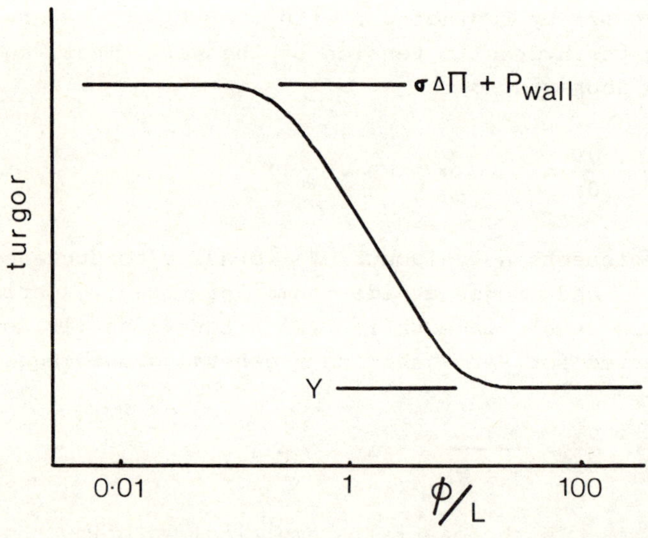

MEASUREMENT OF THE PHYSICAL GROWTH PARAMETERS IN LIVING SYSTEMS

Despite their simplicity and empirical nature the measurement of growth rate parameters has proved to be technically difficult. The best described system so far is that of the growing Nitella internodal cell. Using the earliest version of the pressure probe Green (1968) was able to measure the turgor pressure of these cells directly and continuously during growth. He showed that in this system turgor pressure was indistinguishable from $\Delta\pi$, indicating that $L >> \phi$ and also that $\sigma \simeq 1$. Growth rate of Nitella is thus described adequately by equation (1), from which ϕ could be determined as $2.8 \times 10^{-4}\ s^{-1}.MPa^{-1}$. Subsequently Steudle and Zimmermann (1974) using the non-minaturised pressure probe measured Lp directly for Nitella from which L could be calculated ($2 \times 10^{-2}\ s^{-1}.MPa^{-1}$; Cosgrove, 1981) and thus confirmed to be much greater than ϕ. Determining Y proved not to be so straightforward as it varied under metabolic control (Green et al., 1971) to maintain (P-Y) at about 0.02 MPa regardless of the turgor pressure imposed upon the cell and that following a period of transition, growth rate returned to the previous constant rate despite a different turgor pressure. This "governor" activity was effective for turgor pressure down to about 0.2 MPa - it appears that lower values of Y could not be achieved. Also as turgor approaches this lower limit the degree of compensation gradually diminishes. Green et al., (1971) further indicated that the yield point lowering process is metabolic while the yield raising process is a stress hardening event alone.

Measurement of the reflection coefficients for various solutes with Nitella (Dainty & Ginzburg, 1964; Steudle & Zimmermann, 1974) thus completes the measurement of all the parameters of equation (4) for this organism. In this respect it remains unique.

For higher plants such complete, direct measurements remain to be performed. Cleland (1959) measured the growth rates of Avena coleoptiles bathed in solutions of increasing mannitol concentration to decrease turgor. The preparations behaved according to equation (1), with an apparent yield turgor pressure of about 0.6 MPa. Although Cleland showed that ϕ appeared to be modulated by auxin, no quantitative values were given. A linear dependence of growth rate on turgor pressure has been observed for several tissues as noted above and turgor thresholds demonstrated for stem cells (Lawlor, 1969) and for leaves (Boyer, 1968, 1970; Hsiao *et al*., 1970; Van Volkenburgh & Cleland, 1981).

Since each of the parameters of higher plant growth has the potential of being metabolically modified in the same fashion as yield threshold in *Nitella*, it is not too surprising to find in the literature an almost equal number of observations in which the Lockhart equation apparently does not hold. For example modification of wall extensibility by endogenous growth regulators does not alter growth rate despite constant turgor pressure for *Avena* coleoptiles (Ruesink, 1969); alterations of wall extensibility during the growth of *Phaseolus* leaves makes growth not a direct function of turgor pressure (Van Volkenburgh *et al*., 1985); transverse growth of *Nitella* is not a function of transverse wall extensibility (Metraux & Taiz, 1978).

Unlike *Nitella*, the cells of higher plants are not isolated and not in a stirred medium. Determination of Lp and L is therefore not so straightforward. We shall return to this point in a subsequent section.

MEASUREMENT OF THE RHEOLOGICAL PROPERTIES OF ISOLATED WALLS

Due to the difficulties of measuring ϕ and Y from equation (4) directly on living tissue most work on tissue rheological properties has been performed on "isolated" cell walls, *i.e*. dissected wall segments and loops, in the case of giant algal cells, and tissue segments treated to destroy the protoplast in the case of higher plants (see Cleland, 1981). By definition irreversible wall expansion is central to cell and tissue growth. Such expansion is described as viscoelastic, a list of references for which is given by Cleland (1971). In creep tests following application of a constant stress two extreme types of viscoelastic behaviour may be characterised (Ferry, 1970). At one extreme the material continues to extend, reaching a constant rate of extension. Such material has the properties of a viscoelastic liquid. At the other extreme an equilibrium extension is reached. Such material is termed a viscoelastic solid. These have been alternatively termed viscoelastic and stress hardened flow (Cleland, 1968). Composite materials, such as cell walls, might be expected to possess both properties.

Equation (1), for living tissue, is a description of a viscoelastic liquid. Isolated cell walls, however, do not show this property and behave as viscoelastic solids (Masuda, 1978; Sellen, 1980). In creep experiments on isolated walls extension at constant stress is proportional to log time while that of living cells is proportional to linear time. The difference is due to chemical modification linked to biochemical processes. Cleland (1971) describes this property of the living cell wall as "chemical creep". In physiological terms this is envisaged as a continuous series of events each with the behaviour of a viscoelastic solid (Cleland, 1968). The frequency and magnitude of these events result in the observed liquid flow. Cleland (1983) has recently introduced the concept of capacity for acid wall loosening which appears to be an innate upper limit for this activity during auxin-induced growth. Much work has been performed in attempting to identify the biochemical nature of these wall loosening processes (_e.g._ see Cleland, 1981; Terry _et al_, 1982) which lies beyond the scope of this review. It is, however, important to note that the different growth modulators (auxin, gibberellin, ethylene, light, _etc._) probably employ a range of sites of activity, the relationship of which to the physical growth parameters is poorly understood.

The outcome of this is that _in vitro_ stress/strain analysis cannot provide quantitative data for the empirical parameters of equations (1) and (4). Furthermore, metabolic processes are not the only reason why _in vitro_ rheological studies cannot give quantitative data regarding _in vivo_ extensibility and yield stress. A second difficulty is the necessity of applying uniaxial stress during *in vitro* studies. Cell growth, by contrast, is driven by cell turgor, which is a multiaxial stress. Kamiya *et al*. (1963) showed experimentally that a uniaxial longitudinal stress resulted in 3-4 times more extension than the equivalent stress experienced as a result of turgor pressure. Recently Sellen (1983) has tackled this problem theoretically for *Nitella*, a cell of conveniently simple geometry. An important consideration is the orientation of the stress bearing elements, such as cellulose fibrils, of the wall. This aspect will be considered in the later section on direction of growth.

We are probably some way from a more general or empirical approach although Steudle *et al*. (1982) have attempted to quantify the effect of cellular anisotropy on the elastic component by comparing transverse elasticity with the bulk elastic modulus.

A further drawback of applying uniaxial stress involves the widespread use of "mechanically conditioned" preparations, *i.e*. preparations that have been stressed before measurement (*e.g*. Cleland, 1981). Unlike growing cells, non-conditioned wall preparations have very low or even no yield threshold (Cleland, 1977). This is thought to be due to the fact that uniaxial stresses cause extension along directions (force vectors) that would not normally be stressed during multiaxial stress. During growth the wall loosens in all directions, but only those bonds that are stressed are "used up", the remaining bonds reform in their original positions after a lifetime of an hour or so (Cleland, 1971, 1981). The non-physiological uniaxial stress is thought to extend these remaining loosened bonds and will therefore be a measure of not current or future extensibility of the wall but that of the previous hour or so (Cleland, 1971, 1981).

Several methods have been used to measure the extensibility of isolated walls (see Cleland, 1981; Penny & Penny, 1978). The most widespread technique is probably that of applying a constant rate of increase of strain to the sample while simultaneously measuring stress (e.g. Instron). Other workers (notably Masuda and co-workers) have applied stress relaxation studies following an "instantaneous" extension of the tissue to constant strain (see Masuda, 1978). Such an approach appears to be more readily analysed physically using established rheological methods although theoretical analysis (Fujihara et al., 1978) shows that each of the in vitro techniques gives the same information concerning the rheological properties of the walls. The results of stress relaxation studies may be modelled with a series of Maxwell elements in parallel (Fig. 3) each incorporating an elastic spring and a viscous non-Newtonian dashpot (Ferry, 1970; Masuda, 1978; Dorrington, 1980; Sellen, 1980). Masuda and co-workers (e.g. Yamamoto, Makai & Masuda, 1974) assume stress relaxation to be linear with log time after a lag period, T_0 and express it as the equation -

$$S = b.\log\left(\frac{t + T_m}{t + T_0}\right) + C \tag{5}$$

Fig. 3. Mechanical model comprising of springs and "dashpots" (hydraulic dampers) used to describe the viscoelastic properties of cell walls. A, Maxwell element; B, generalised model (after Masuda, 1978).

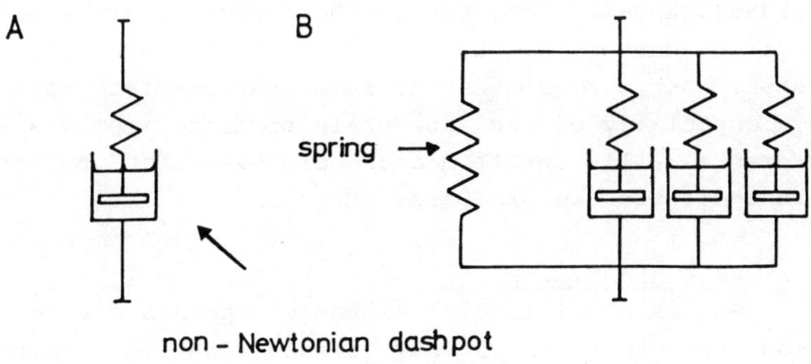

This model introduces the concept of a relaxation spectrum (Ferry, 1970; Wainwright _et_ _al_., 1976) which describes viscoelastic relaxation as a series of discrete events each with its own time constant. This explains that "instantaneous" irreversible extension and creep are essentially the same phenomenon, the former representing creep that occurs in a very short time period (Cleland, 1971). Cell walls vary enormously in the relative shapes of their relaxation spectra, _e.g_. for _Nitella_ rapid relaxations are most significant, while for wood very much longer time scale extension predominates the spectrum (Dorrington, 1980). The existence of the spectrum explains the strict requirement for reproduceable conditions during mechanical extension methods, _e.g_. strain rate during Instron experiments. Different time courses of experiment would result in different portions of the spectrum being studied. Different portions might well possess very different properties, _e.g_. response to plant growth substances.

It is hardly surprising that _in_ _vitro_ studies cannot give quantitative values for the parameters in equation (1). Do such studies, therefore, contribute at all to out understanding of ϕ and Y? The answer must be yes, if only qualititively, since in many studies the various parameters measured _in_ _vitro_ have been correlated qualititively with growth rate modulation (see Cleland, 1971, 1983; Masuda, 1978; Preston, 1974; Van Volkenburg _et_ _al_., 1985). Equally importantly rheological studies provide information concerning the molecular basis of cell wall extension. As our understanding of the chemistry of cell walls, especially of the interchain bonding, improves these measurements will enable macroscopic events to be traced back to their microscopic bases.

WALL ELASTICITY

Reversible (elastic) expansion appears not to play an important role in steady state kinetics, although Cosgrove (1981) has described the influence of wall elasticity on the duration of adjustment following a change in any of the steady state parameters of equation (4). Such transitions may well be common in a natural environment, *e.g.* during diurnal changes of the status of the plant. It is quite possible, however that elastic properties may be qualitative indicators of extensibility parameters in some cases. Cleland (1967), for example, indicated that elastic compliance correlated with growth rate at a level equal to plastic compliance, while Masuda and co-workers (Masuda, 1968, 1969; Masuda *et al.*, 1970) claimed that elastic compliance correlated better with growth than did plastic compliance. Cleland (1971) pointed out that these effects were very small, although he did indicate the possibility that the stretching of wall components as a result of elastic properties may be the crucial factor in the yield threshold rather than pressure *per se*.

Zimmermann & Steudle (1978) have demonstrated the turgor and volume dependence of volumetric elastic modulus for a range of plant tissues. Thus it may prove a useful indirect indicator of endogenous growth limiting factors based on wall viscoelasticity. This may be of particular relevance to the factors that determine mature cell size, an aspect that is far from being understood. However, the relationship between the elastic and viscoelastic properties of cell walls has yet to be elucidated.

WALL AND TISSUE HYDRAULIC CONDUCTIVITY

The relevance of the ϕ/L ratio has already been discussed. For *Nitella* L is not limiting to growth. Use of the pressure probe has indicated that Lp values for a range of individual plant cells are of a similar order of magnitude to those of *Nitella* (Zimmermann & Steudle, 1978; Tomos & Zimmermann, 1982; Steudle, Tyerman & Wendler, 1983). For cells with walls of similar extensibility to *Nitella* growing in isolation, therefore, L is unlikely to be a limiting factor and growth will approximate to equation (1). The cells of growing leaves and other tissues, however, are *not* isolated and L will be influenced by the hydraulic conductivity of the entire pathway from soil to cell vacuole. This has been considered to be growth limiting by several authors, e.g. Boyer (1977), Boyer & Wu (1978) Cutler et al., (1980), Tyree & Jarvis (1982), Westgate & Boyer (1984).

Estimation of the conductance of this pathway is hampered by our lack of total understanding of the path followed by water. Three potential pathways of liquid water flow through tissues have been envisaged (Molz & Ferrier, 1982). These are the vacuole to vacuole pathway, described quantitively in isolation by Philip (1958), the apoplasmic pathway through the continuum of cell walls (Scott & Priestley, 1928) and the symplasmic pathway including bulk transport of water through the plasmodesmata. This last pathway, although extensively discussed for roots (e.g. Weatherley, 1982), is as yet poorly characterised for leaves. These pathways are not exclusive and will occur in parallel and in series (see Chapter 2).

Resistance to water flow in the xylem of small herbaceous plants is generally considered to be rather low in comparison to the other portions of the pathway. A general treatment of the hydraulic conductivity of the xylem of trees is given in Zimmermann & Brown (1971). Radial movement of water across the root has been studied by several workers (see Weatherley 1982; Jones et al., 1983; Steudle & Jeschke, 1983), but has also been considered to be of relatively low resistance on the pathway to the growing tissue (Boyer & Wu, 1978; Molz & Boyer, 1978; Cosgrove & Cleland, 1983b). The major potential influence on the value of L for the leaf cell growth is therefore likely to be the path of water through the leaf tissue itself (Boyer, 1977; Westgate & Boyer, 1984).

Attempts to quantify the hydraulic conductance and diffusivity (see Molz & Ikenberry, 1974) of leaf tissue requires the measurement of each of the three pathways. Using an electrical circuit analog model, Molz and Ferrier (1982) detail the parameters required to perform this description and have attempted to glean values for them from the literature. They admit, however, that it is "an experimental challenge that has yet to be met fully".

Resistance to water flow within a tissue will lead to the development of water potential gradients the magnitude of which would be a direct measurement of L in equation (4). Molz and Ferrier (1982) use the available data to derive theoretical profiles for these gradients and illustrate with specific reference to the model systems of growing soybean hypocotyl (Molz & Boyer, 1978) and primary maize root (Silk & Wagner, 1980). The conclusion is that significant water potential gradients are set up across these tissues, the magnitude of which depends on the diffusivity of the tissue. Gradients of about 0.2 to 0.3 MPa across the cortex of growing soybean hypocotyl would be consistent both with the theory and with the observation that the average water potential of the entire cortex was 0.17 to 0.21 MPa lower than the root medium in growing intact tissue (Molz & Boyer, 1978). Similar gradients are proposed for growing root tissue (Silk & Wagner, 1980) and from the various growing zones, including leaves, of maize (Westgate & Boyer, 1984).

In contrast Cosgrove and Cleland (1983b) have recently questioned the existence of a significant water potential gradient across the cortex of growing pea epicotyls. These workers suggest that the observed low water potential measured from expressed sap is due to the presence of apoplasmic solutes (Cosgrove & Cleland, 1983b). Using the pressure probe they have decisively demonstrated that in this tissue there is no significant difference in turgor pressure in the individual cells across the cortex. Their evidence for no osmotic potential gradient is probably less convincing and requires further investigation. (They show that both parts of assymmetrically split tissue have the same osmotic potential.) The lack of a turgor pressure gradient across the tissue is reassuring as such a gradient would demand a compensatory gradient in the rheological properties to prevent a "telescoping" of the tissue during growth. If a water potential gradient does exist, it must be due to the osmotic component alone, and parallel gradients of osmotic potential must exist in both the apoplast and intracellular components. Presumably this balance would have to be actively regulated by osmoregulation mechanisms. Westgate and Boyer (1984) have shown that osmotic regulation is important in maintaining a constant water potential gradient across the growing leaf during diurnal variation of xylem water potential due to transpiration. Such a mechanism would provide an important regulation role for the apoplast solutes (Cosgrove & Cleland, 1983a) similar to that suggested for beetroot tissue (Leigh & Tomos, 1983). The plasmodesmata may play an important role in maintaining the constant hydrostatic pressure across an entire tissue.

The water potential gradients due to growth will evidently be integrated with gradients due to transpiration. This aspect has recently been discussed theoretically by Fiscus et al. (1983) and for maize by Westgate and Boyer (1984).

Tissue architecture must play a very important role in leaf, or other tissue, water relations. The apparent role of the Casparian strip in roots to regulate the roles of the parallel hydraulic pathways is an excellent example of the importance of such detail. In this respect growing epicotyls and roots may be very poor models for leaf growth. Mathias (1983) has recently presented a theoretical description of extracellular tortuosity which may be of significant value in this field. A general model of leaf transport at the entire organ level has been presented by Tyvand (1982).

A direct method of estimating L, or rather its importance as a limitation to growth, would be to investigate turgor as a function of growth. By using the dependence of P on the ϕ/L ratio (Cosgrove, 1983) a decrease of turgor on initiation of growth might indicate that L is of similar magnitude to ϕ. Unfortunately the third important parameter of interest, osmoregulation, will also influence the response of turgor to growth.

THE MAINTENANCE OF OSMOTIC POTENTIAL DURING GROWTH

As water enters the expanding cells the sap osmotic

potential would rise (become less negative) unless solute accumulation occurred simultaneously. This solute accumulation could be due either to intracellular synthesis or to uptake of extracellular solutes. Early work by Cleland (1959) indicated that for continuous growth of excised Avena coleoptiles an external source of sucrose was essential. McNeil (1976) has indicated that in intact plants phloem transport of sucrose may play an important role in the osmoregulation of the growing zone of plants.

Table 2 illustrates the expected effects on growing cells of L and/or osmoregulation having a limiting effect on growth. Both lower turgor pressure, the former as the cells cease to be in osmotic equilibrium with their water supply and the latter due to the lack of compensation of the dilution of the vacuolar sap.

Table 2. The influence of L, ϕ and solute fluxes on water potential equilibrium, turgor and cellular osmotic potential in growing cells.

	Water potential	Turgor pressure	Osmotic potential
a) $L>>\phi$			
Solute flux			
(i) not limiting	equilibrium	constant	constant
(ii) limiting	equilibrium	lowered	raised
b) $L<=\phi$			
Solute flux			
(i) not limiting	non-equilibrium	lowered	constant
(ii) limiting	non-equilibrium	lowered	raised

Following the results of Ordin et al. (1956) for Avena coleoptiles solute accumulation during growth has generally been considered to be non-rate limiting, and that intracellular osmotic potential in equations (3) and (4) remains constant.

Plant cells are certainly capable of regulating total osmotic potential and the relative composition of their vacuoles. Most studies, however, have followed this over a time scale measured in days (Hsiao, 1973; MacClachlan et al., 1970; Pitman et al., 1971; Gracean & Oh, 1972). How readily, therefore, can growing cells maintain a low osmotic potential?

It would appear from the literature that some tissues maintain osmotic potential during growth while others do not. For example no significant change in osmotic potential was observed following growth regulator-stimulated growth for Avena coleoptile (Ordin et al., 1956; Stevenson & Cleland, 1981, 1982), cucumber hypocotyl, (Cohen & Atsmon, 1972) or Pisum internodes (Cosgrove & Cleland, 1983b). On the other hand Katsumi and Kazama (1978) have demonstrated that in cucumber osmotic potential actually decreases following growth stimulation by GA_3. (These authors suggest that GA_3 stimulation of growth in this tissue is due to decreased osmotic potential since wall extensibility is not affected. A similar result was noted for tomato stem (Van den Ende & Koorneef, 1960).

For gibberellin stimulated growth of lettuce sections, on the other hand, osmotic potential increases in parallel with growth (Stuart & Jones, 1977, 1978) unless 10mM KCl is present in the bathing medium when osmotic potential remains unchanged. In this system it is interesting to note that incubation in 10mM KCl alone decreases osmotic potential (and presumably increases turgor) without altering the rate of growth. Recently light stimulated growth of Phaseolus leaves has been shown also to be accompanied by an increase in osmotic potential (Van Volkenburgh & Cleland, 1981; Davies & Van Volkenburgh, 1983). In this system also growth rates do not follow changes in turgor since an increased water potential (and presumably turgor) is not followed by a decrease in growth rate (Van Volkenburgh et al., 1985). These last examples serve as a reminder that several of the physical parameters will be changing simultaneously on stimulation, e.g. cell extensibility (ϕ) as well as π_i in growing Phaseolus leaves (Van Volkenburgh et al., 1984).

In none of these studies is it possible to determine whether a limiting L also influences turgor (and osmotic potential) since the method used to determine turgor (with the exception of those using the micro pressure-probe; Cosgrove & Cleland, 1983b) makes the implicit assumption that the cells are in osmotic equilibrium.

It would appear, therefore, that osmoregulation may or may not be limiting to growth and that it may well be a control point for the influence of growth regulators, along with other parameters. The regulation may be either at the level of the individual growing cell, i.e. membrane transport or metabolism, or at the level of the tissue, e.g. phloem transport (McNeil, 1976). Here again tissue architecture is likely to be of importance, with the distance and transport capacity between phloem and expanding cells playing an important role.

Interaction between growth and solute accumulation will be at its most critical in the case of cells with very low osmotic potentials, e.g. halophytes. Greenway and Munns (1983) have recently reviewed their interrelationship. They show that growth at constant π_i must follow an equation of the form:-

$$\frac{dV}{dt} = \left(\frac{\text{Net solute uptake + Net solute synthesis}}{-\pi_i}\right)\frac{RT}{1} \qquad (6)$$

i.e. that growth is a function of π_i and that it will ultimately be limited by rate of solute synthesis and uptake. These authors suggest that for halophytes maximum growth will be determined by Cl^- and Na^+ uptake capacity and show that one of them at least will involve an active carrier. Jeschke et al. (1983) further suggests that at high salinity K^+ may be limiting.

Few data are available to quantify these parameters, although Greenway and Munns (1983) have illustrated the order of magnitude of fluxes required by calculating that maize roots growing at 100mM NaCl would require Na^+ and Cl^- flux rate of about 50×10^{-9} mol m^{-2} s^{-1}

to maintain the unstressed growth rate observed (Erickson & Sax, 1956). Cram (1973) and Pitman (1976), however, have estimated Cl^- influx across the tonoplast of cortical maize root cells exposed to 100mM Cl^- to be only 18×10^{-9} mol m^{-2} s^{-1}. Greenway and Munns (1983) conclude that for this salt sensitive species Cl^- transport would be inadequate above external concentration of about 50mM. Halophytes, however, are probably capable of more rapid Cl^- and Na^+ uptake. Indeed much higher rates have been observed for marine microalgae in seawater (500mM Cl^-). Raven (1976) estimates tonoplast flux values at $900\text{-}5600 \times 10^{-9}$ mol m^{-2} s^{-1} for Cl^-; $40\text{-}120 \times 10^{-9}$ mol m^{-2} s^{-1} for Na^+ and $1000 - 95{,}000 \times 10^{-9}$ mol m^{-2} s^{-1} for K^+.

Greenway and Munns (1983) also consider whether transport capacity or the metabolic energy required to drive uptake is likely to be limiting, and come to the conclusion that at present insufficient data are available to choose. Such considerations are beyond the scope of the review.

Osmoregulation may be important not only in halophytes with low water potentials but also in plants growing under varying environmental conditions, e.g. water potential variation due to transpiration (Cleland, 1977). For example, Michelena and Boyer (1982) and Westgate and Boyer (1984) have shown that osmoregulation allows turgor maintenance during the day in growing cells of maize leaves. It is important to note, however, that despite this regulation, growth stops as water potentials fall.

DIRECTION OF CELLULAR GROWTH

The direction of expansion of most growing cells is probably determined by the orientation of the cellulose fibrils and other polysaccharide components of the wall (see Cleland, 1971; Preston, 1974; Morikawa, Kitamura & Senda, 1978; Roland & Vian, 1979). The ultrastructural basis of development is still a matter of dispute with the multinet growth hypothesis of Roelofsen and Houwink (1953) being challenged by more dynamic models such as the ordered growth hypothesis (Roland, Vian & Reis, 1975, 1977; Roland & Vian,

1979; Takeda & Shibaoka, 1981a, 1981b). Here again different cell types may use different extension mechanisms.

From a physical point of view, the best understood system is once again the growing Nitella cell. For such cylindrical cells it can be shown that tangential stress will be twice that experienced longitudinally (Nobel, 1974).

$$S_L = rP/d \quad (7)$$

$$S_T = rP/2d \quad (8)$$

Therefore, to maintain the preferable longitudinal growth, as observed in Nitella, an anisotropy of extensibility, and/or threshold yield stress must occur. Probine and Preston (1962) were able to demonstrate such an anisotropy in both the elastic and viscoelastic extensibility properties of isolated cells. The ratio of transverse to longitudinal elastic modulus was about 5 for growing cells and diminished to 2 for mature cells. Metraux and Taiz (1978) subsequently demonstrated that the threshold yield stress in the transverse direction is twice that in the longitudinal direction, and confirmed the anisotropy of extensibility, finding the ratio of transverse viscoelastic extensibility to longitudinal decreasing from 4 in growing cell to about 2 as the cells grow older. Qualitatively, therefore, both Y and extensibility are consistent with growth form. Although quantitatively, here again, there are inconsistences, especially the observation that despite transverse extensibility remaining constant throughout growth, the tranverse growth rate is not constant. The explanation must lie, as noted in a previous section, in the fact that cell expansion is a biochemical as well as a physical process. Metraux and Taiz (1979) subsequently characterised other properties of a physiological nature that distinguish the two directions of extension.

Metraux, Richmond and Taiz (1980) have demonstrated the influence on growth of endogenous pH gradients along the wall of *Nitella*. Similar gradients could well be responsible for "bulges" and other outlines of irregularly shaped cells. Such gradients might be more stable in higher plants where there would be no bathing solution, as with *Nitella*, to remove the gradients.

The mechanical axis of *Nitella* cell wall does not correspond exactly with the cell axis, but may lie as much as 10° from it (Probine, 1963). The effect of this is to "twist" the cell as it extends. Spiral growth has been described for the sporangiophore of *Phycomyces* (see Preston, 1974), although a spiral fibrillar orientation may be common to all plant cells (Robards, 1974). Sellen (1980) considers that the twisting effect will be inhibited by the presence of neighbouring cells. Preston (1974), however, suggests that spiral growth may play an important role in such phenomena as the spiral grain of conifers, the movement of elaters in liverworts, the spiraling of tendrils and the nutation of stem apices.

Sellen (1983) has recently presented a theoretical analysis of the responses of mechanically anisotrophic cylindrical cells, such as *Nitella*, to multiaxial stress.

CONCLUSIONS

Although we are still a long way from fully understanding the physical parameters that determine the patterns of growth of leaf cells, it seems likely that many of the properties of other "model" tissues and cells will prove equally relevant to leaves. For example such properties as acid wall loosening of leaf cells (Van Volkenburgh & Cleland, 1980) presumably have the same basis as that found in *Avena* coleoptiles *etc*. Leaf cells have been shown to have yielded stress thresholds *in vivo* (Van Volkenburgh & Cleland, 1981) and their *in vitro* viscoelastic properties are also comparable to the "model" systems in terms of extensibility (Van Volkenburgh *et al*., 1985).

The water relations properties of individual leaf cells can now be readily studied with the pressure probe (Zimmermann & Steudle, 1978). However, our understanding of the water relations of entire tissues is still far from complete although theoretical models such as that of Molz and Ferrier (1982) point the direction for future work. We must, however, pay attention to the specific anatomy of the tissue under study before we can usefully use such models. In the study of root hydraulic conductivity it has been possible to compare the model predictions against whole tissue measurements (Jones *et al.*, 1983; Steudle & Jeschke, 1983) although in this case some confusion has been caused by over-simplifying the descriptions of the tissue anatomy. In leaves it is not so easy to compare model predictions (*e.g.* Tomos *et al.*, 1981) with any whole leaf parameters. The supply of osmotic solutes to maintain low osmotic potentials in growing cells may be the physical growth parameter that is most readily governed by biochemical and physiological processes. Of necessity solute uptake must be closely coupled with water uptake, and although purely passive fluxes of the type described by Goldman (1943) appear to be sufficient to load the individual cell with most of its requirements some active processes, such as organic solute synthesis or solute pumping, must be involved and be under tight feedback control.

The growth of each cell is governed by a number of independent and partially dependent variables. Such observations might suggest that the use of the Lockhart equation is not very helpful in practice. However, it does provide an empirical basis for the parameters. It is important, however, to realise that each of these parameters need not be constant and that several of them may be modified simultaneously.

ACKNOWLEDGEMENTS

I am grateful for the receipt of reprints, preprints and helpful suggestions from Professor R.G. Wyn Jones and Drs D. Cosgrove, E. Steudle, J.L. Stoddart and E. Van Volkenburgh. The review was written during the term of an AFRC fellowship.

REFERENCES

Boyer, J.S. (1968). Relationship of water potential to growth in leaves. Plant Physiol., 43, 1056-62.

Boyer, J.S. (1970). Leaf enlargement and metabolic rates in corn, soybean and sunflower at various leaf water potentials. Plant Physiol., 46, 233-35.

Boyer, J.S. (1977). Regulation of water movement in whole plants. In Integration of Activity in the Higher Plant, ed. D.H. Jennings. Soc. Exptl. Biol. Symp., 31, 455-70.

Boyer, J.S. & Wu, G. (1978). Auxin increases the hydraulic conductivity of auxin sensitive hypocotyl tissue. Planta, 139, 227-37.

Bunce, J.A. (1977). Leaf elongation in relation to leaf water potential in soybean. J. Exp. Bot., 28, 156-61.

Cleland, R.E. (1959). Effect of osmotic concentration on auxin action and on irreversible and reversible expansion of *Avena* coleoptiles. Physiol. Plant, 12, 809-25.

Cleland, R.E. (1967). Extensibility of isolated cells walls" measurement and changes during cell elongation. Planta, 74, 197-209.

Cleland, R.E. (1968). Wall extensibility and the mechanism of auxin-induced cell elongation. In Biochemistry and Physiology of Plant Growth Substances, eds. F. Wightman & G. Setterfield, pp. 613-42. Ottawa: Runge Press.

Cleland, R.E. (1971). Cell wall extension. Ann. Rev. Plant Physiol., 22, 197-222.

Cleland, R.E. (1977). The control of cell enlargement. In Integration of Activity in the Higher Plant, ed. D.H. Jennings. Soc. Exptl. Biol. Symp., 31, 101-15.

Cleland, R.E. (1981). Wall extensibility: hormones and wall extension. In Plant Carbohydrates II. Extracellular Carbohydrates, eds. W. Tanner & F.A. Loewus. Encycl. Plant Physiol., 13B, 255-73.

Cleland, R.E. (1983). The capacity for acid-induced wall loosening as a factor in the control of *Avena* coleoptile cell elongation. J. Exp. Bot., 34, 676-80.

Cohen, D. & Atsmon, D. (1972). Measurements of water potentials and hormone transport associated with the growth of cucumber hypocotyls. In Plant Growth Substances 1970, ed. D.J. Carr. pp. 23-9. Berlin: Springer-Verlag.

Cosgrove, D.J. (1981). Analysis of the dynamic and steady-state responses of growth rate and turgor pressure to changes in cell parameters. Plant Physiol., 68, 1439-46.

Cosgrove, D.J. (1983). Photocontrol of extension growth: a

biophysical approach. Phil. Trans. R. Soc. Lond. B., 303, 453-65.
Cosgrove, D.J. & Cleland, R.E. (1983a). Solutes in the free space of growing stem tissues. Plant Physiol., 72, 326-31.
Cosgrove, D.J. & Cleland, R.E. (1983b). Osmotic properties of pea internodes in relation to growth and auxin action. Plant Physiol., 73, 332-38.
Cram, W.J. (1973). Chloride fluxes in cells of the isolated root cortex of *Zea mays*. Aust. J. Biol. Sci., 26, 757-79.
Cutler, J.M., Steponkus, P.L., Wach, M.J. & Shahan, K.W. (1980). Dynamic aspects and enhancement of leaf elongation in rice. Plant Physiol., 66, 147-52.
Dainty, J. & Ginzburg, B.Z. (1964). The reflection coefficient of plant cell membranes for certain solutes. Biochim. Biophys. Acta, 79, 129-37.
Davies, W.J. & Van Volkenburgh, E. (1983). The influence of water deficit on the factors controlling the daily pattern of growth of *Phaseolus* trifoliates. J. Exp. Bot., 34, 987-99.
Dorrington, K.L. (1980). The theory of viscoelasticity in biomaterials. *In* the Mechanical Properties of Biological Materials, eds. J.F.V. Vincent & J.D. Currey. Soc. Exptl. Biol. Symp., 34, 289-314.
Erickson, R.O. & Sax, K.B. (1956). Elemental growth rate of the primary root of *Zea mays*. Proc. Amer. Phil. Soc., 100, 487-98.
Ferry, J.D. (1970). Viscoelastic Properties of Polymers. (Second edition). Chichester: John Wiley.
Fiscus, E.L., Klute, A. & Kauffmann, M.R. (1983). An interpretation of some whole plant water transport phenomena. Plant Physiol., 71, 810-17.
Fujihara, S., Yamamoto, R. & Masuda, Y. (1978). Viscoelastic properties of plant cell walls. - 3. Hysteresis loop in the stress-strain curve at constant strain rate. Biorheology, 15, 87-9.
Goldman, D.E. (1943). Potential, impedance and rectification in membranes. J. Gen. Physiol., 27, 37-60.
Greacen, E.L. & Oh, J.S. (1972). Physics of root growth. Nature, 235, 24-5.
Green, P.B. (1968). Growth physics in *Nitella*: A method for continuous *in vivo* analysis of extensibility based on a micro-manometer technique for turgor pressure. Plant Physiol., 43, 1169-84.
Green, P.B. (1969). Cell morphogenesis. Ann. Rev. Plant Physiol., 20, 365-94.
Green, P.B., Erickson, R.O. & Buggy, J. (1971). Metabolic and physical control of cell elongation rate. *In vivo* studies in *Nitella*. Plant Physiol., 47, 423-30.
Green, P.B., Erickson, R.O. & Richmond, P.A. (1970). On the

physical basis of cell morphogenesis. Ann. N.Y. Acad. Sci., 175, 712-31.
Greenway, H. & Munns, R. (1983). Interactions between growth, uptake of Cl and Na and water relations of plants in saline environments. II. Highly vacuolated cells. Plant Cell and Environ., 6, 575-89.
Grenetz, P.S. & List, A. Jr. (1973). A model for predicting growth responses in plants to changes in external water potential: Zea mays primary roots. J. Theor. Biol., 39, 29-45.
Hettiaratchi, D.R.P. & O'Callaghan, J.R. (1974). A membrane model of plant cell extension. J. Theor. Biol., 45, 459-65.
Hsiao, T.C. (1973). Plant responses to water stress. Ann. Rev. Plant Physiol., 24, 519-70.
Hsiao, T.C., Acevedo, E., Fereres, E. & Henderson, D.W. (1976). Stress metabolism. Water stress, growth and osmotic adjustment. Phil. Trans. R. Soc. Lond. B., 273, 479-500.
Hsiao, T.C., Acevedo, E. & Henderson, D.W. (1970). Maize leaf elongation: Continuous measurement and close dependence on plant water status. Science, 168, 590-1.
Jeschke, W.D., Stelter, W., Reising, B. & Behl, R. (1983). Vacuolar Na^+/K^+ exchange, its occurence in root cells of Hordeum, Atriplex and Zea and its significance for K^+/Na^+ discrimination in roots. J. Exp. Bot., 34, 964-79.
Jones, H., Tomos, A.D., Leigh, R.A. & Wyn Jones, R.G. (1983). Water-relation parameters of epidermal and cortical cells in the primary root of Triticum aestivum L. Planta, 158, 230-6.
Jones, R.L. (1982). Gibberellin control of cell elongation. In Plant Growth Substances 1982, ed. P.F. Wareing. pp. 121-30. New York: Academic Press.
Kamiya, N., Tazawa, M. & Taketa, T. (1963). The relation of turgor pressure to cell volume in Nitella with special reference to the mechanical properties of the cell wall. Protoplasma, 57, 501-21.
Katchalsky, A. & Curran, P.F. (1965). Non-equilibrium thermodynamics in biophysics. Cambridge, Mass: Harvard University Press.
Katsumi, M. & Kazama, H. (1978). Gibberellin control of cell elongation in cucumber hypocotyl sections. Bot. Mag. Tokyo Special Issue, 1, 141-58.
Kirkham, M.B., Gardner, W.R. & Gerloff, G.C. (1972). Regulation of cell division and cell enlargement by turgor pressure. Plant Physiol., 49, 961-2.
Lawlor, D.W. (1969). Plant growth in polyethylene glycol solutions in relation to the osmotic potential of the root medium and the leaf water balance. J. Exp.

Bot., 20, 895-911.

Leigh, R.A. & Tomos, A.D (1983). An attempt to use isolated vacuoles to determine the distribution of sodium and potassium in cells of storage roots of red beet (*Beta vulgaris* L.). Planta, 159, 469-75.

Lockhart, J.A. (1965). An analysis of irreversible plant cell elongation. J. Theor. Biol., 8, 264-75.

Maksymowych, R. (1973). Analysis of Leaf Development. Cambridge: Cambridge University Press.

Masuda, Y. (1968). Role of the cell-wall degrading enzymes in cell-wall loosening in oat coleoptiles. Planta, 83, 171-84.

Masuda, Y. (1969). Auxin-induced cell expansion in relation to cell wall extensibility. Plant Cell. Physiol., 10, 1-9.

Masuda, Y. (1978). Auxin-induced cell wall loosening. Bot. Mag. Tokyo Special Issue, 1, 103-23.

Masuda, Y., Pjon, C-J. & Furuya, M. (1970). Phytochrome action in *Oryza sativa* L. 5 - Effects of decapitation and red and far-red light on cell wall extensibility. Planta, 90, 236-42.

Mathias, R.T. (1983). Effect of tortuous extracellular pathways on resistance measurements. Biophys. J., 42, 55-9.

MacClachlan, G.A., Datko, A.H., Rollit, J. & Stokes, E. (1970). Sugar levels in the pea epictoyl: regulation by invertase and sucrose synthetase. Phytochem. 9, 1023-30.

McNeil, D.L. (1976). The basis of osmotic pressure maintenance during expansion growth in *Helianthus annuus* hypocotyls. Aust. J. Plant Physiol, 3, 311-24.

Metraux, J.P. & Taiz, L. (1978). Transverse viscoelastic extension in *Nitella*. I. Relation to growth rate. Plant Physiol., 61, 135-8.

Metraux, J.P. & Taiz, L. (1979). Transverse viscoelastic extension in *Nitella*. II. Effect of acid and ions. Plant Physiol. 63, 657-9.

Metraux, J.P., Richmond, P.A. & Taiz, L. (1980). Control of cell elongation in *Nitella* by endogenous cell wall pH gradients. Plant Physiol., 65, 204-10.

Michelena, V.A. & Boyer, J.S. (1982). Complete turgor maintenance at low water potentials in the elongation region of maize leaves. Plant Physiol., 69, 1145-9.

Molz, F.J. & Boyer, J.S. (1978). Growth-induced water potentials in plant cells and tissues. Plant Physiol., 62, 423-9.

Molz, F.J. & Ferrier, J.M. (1982). Mathematical treatment of water movement in plant cells and tissue; a review. Plant Cell and Environ., 5, 191-206.

Molz, F.J. & Ikenberry, E. (1974). Water transport through

plant cells and cell walls: theoretical development. Soil Sci. Soc. Amer. Proc., 38, 699-704.

Morikawa, H., Kitamura, S. & Senda, M. (1978). Effect of auxin on changes in the orientated structure of wall polysaccharides in response to mechanical extension in oat coleoptile cell walls. Plant Cell Physiol., 19, 1553-6.

Nobel, P.S. (1974). Introduction to Biophysical Plant Physiology. San Fransisco: Freeman.

Nutting, P.G. (1921). A new general law of deformation. J. Franklin Inst., 191, 679-86.

Ordin, L., Applewhite, T.H. & Bonner, J. (1956). Auxin induced water uptake by *Avena* coleoptile sections. Plant Physiol., 31, 44-53.

Penny, P. & Penny, D. (1978). Rapid responses to phytohormones. *In* Phytohormones and Related Compounds - A Comprehensive Treatise. Vol. 2., eds. D.S. Letham, P.B. Goodwin and T.J.V. Higgins, pp. 537-97. North Holland: Elsevier.

Pfeffer, W. (1881). Pflanzenphysiologie Bd. 2. Tübingen.

Philip, J.R. (1958). Osmosis and diffusion in tissue: Half times and internal gradients. Plant Physiol., 33, 275-8.

Pitman, M.G. (1976). Ion uptake by plant roots. *In* Transport in Plants I, eds., U. Luttge & M.G. Pitman. Encycl. Plant Physiol., 28, 95-128.

Pitman, M.G., Mowat, J. & Nair, H. (1971). Interaction of processes for accumulation of salt and sugar in barley plants. Aust. J. Biol. Sci., 24, 619-31.

Preston, R.D. (1974). The Physical Biology of Plant Cell Walls. London: Chapman and Hall.

Probine, M.C. (1963). Cell growth and the structure and mechanical properties of the wall of the internodal cells of *Nitella opaca*. - 3. Spiral growth and wall structure. J. Exp. Bot., 14, 101-13.

Probine, M.C. & Preston, R.D. (1962). Cell growth and the structure and mechanical properties of the wall in internodal cells of *Nitella opaca*. II Mechanical properties of the walls. J. Exp. Bot., 13, 111-27.

Quaite, E., Parker, R.E. & Steer, M.W. (1983). Plant cell extension: structural implications for the origin of the plasma membrane. Plant Cell. Environ., 6, 429-32.

Raven, J.A. (1976). Transport in algal cells. *In* Transport in Plants II., eds. U. Luttge and M.G. Pitman. Encycl. Plant. Physiol., 2A, 129-88.

Ray, P.M., Green, P.B. & Cleland, R. (1972). Role of turgor in plant cell growth. Nature, 239, 163-4.

Robards, A.W. (1974). Dynamic aspects of plant ultrastructure. London: McGraw Hill.

Roelofsen, P.A. & Houwink, A.L. (1953). Architecture and

growth of the primary cell wall in some plant hairs and in the _Phycomyces_ sporangiophore. Acta. Bot. Neerl., _2_, 218-25.
Roland, J.C., Vian, B. & Reis, D. (1975). Observations with cytochemistry and ultracryotomy on the fine structure of the expanding walls in actively elongating plant cells. J. Cell Sci., _19_, 239-59.
Roland, J.C., Vian, B. & Reis, D. (1977). Further observations on cell wall morphogenesis and polysaccharide arrangements during plant growth. Protoplasma, _91_, 125-41.
Roland, J.C. & Vian, B. (1979). The wall of the growing plant cell: Its three-dimensional structure. Int. Rev. Cytol., _61_, 129-66.
Ruesink, A.W. (1969). Polysaccharides and the control of cell wall elongation. Planta, _89_, 95-107.
Scott, L.I. & Priestley, J.H. (1928). The root as an absorbing organ. I. A reconsideration of the entry of water and salts in the absorbing region. New Phytol., _27_, 125-40.
Sellen, D.B. (1980). The mechanical properties of plant cell walls. _In_ The Mechanical Properties of Biological Materials, eds. J.F.V. Vincent & J.D. Currey. Soc. Exptl. Biol. Symp., _34_, 315-29.
Sellen, D.B. (1983). The response of mechanically anistropic cylindrical cells to multiaxial stress. J. Exp. Bot., _34_, 681-7.
Silk, M.W.K. & Erickson, R.O. (1979). Kinematics of plant growth. J. Theor. Biol., _76_, 481-501.
Silk, W.K. & Wagner, K.K. (1980). Growth sustaining water potential distribution in the primary corn root. Plant Physiol., _66_, 859-63.
Squire, G.R., Black, C.R. & Ong, C.K. (1983). Response to saturation deficit in leaf extension in a stand of pearl millet (_Pennisetum typhoides_ S & H). II. J. Exp. Bot., _34_, 856-65.
Steudle, E. & Zimmermann, U. (1974). Determination of the hydraulic conductivity and the reflection coefficients in _Nitella flexilis_ by means of direct cell-turgor pressure measurements. Biochem. Biophys. Acta, _332_, 399-412.
Steudle, E., Ferrier, J.M. & Dainty, J. (1982). Measurement of the volumetric and transverse elastic extensibilities of _Chara corallina_ internodes combining external force and pressure probe technique. Can. J. Bot., _60_, 1503-11.
Steudle, E. & Jeschke, W.D. (1983). Water transport in barley roots. Measurement of root pressure and hydraulic conductivity of roots in parallel with turgor and hydraulic conductivity of root cells. Planta, _158_, 237-48.
Steudle, E., Tyerman, S.D. & Wendler, S. (1983). Water

relations of plant cells. In Effects of Stress on Photosynthesis, eds., R. Marcelle, H. Clijsters & M. van Poucke. pp. 95-109. The Hague: Martinus Nijhoff/Junk.

Stevenson, T.T. & Cleland, R.E. (1981). Osmoregulation in the *Avena* coleoptile in relation to auxin and growth. Plant Physiol., 67, 749-53.

Stevenson, T.T. & Cleland, R.E. (1982). Osmoregulation in the *Avena* coleoptile. Control of solute uptake in peeled sections. Plant Physiol., 69, 292-5.

Stuart, D.A. & Jones, R.L. (1977). Roles of extensibility and turgor in gibberellin- and dark-stimulated growth. Plant Physiol., 59, 61-8.

Stuart, D.A. & Jones, R.L. (1978). Role of cation and anion uptake in salt-stimulated elongation of lettuce hypocotyl sections. Plant Physiol., 61, 180-3.

Takeda, K. & Shibaoka, H. (1981a). Changes in microfibril arrangement on the inner surface of the epidermal cell wall in the epicotyl of *Vigna angularis* Ohwi et Ohashi during cell growth. Planta, 151, 385-92.

Takeda, K. & Shibaoka, H. (1981b). Effects of gibberellin and colchicine on microfibrillar arrangement in epidermal cell walls of *Vigna angularis* Ohwi et Ohashi epicotyls. Planta, 151, 393-8.

Terry, M.E., McGrawe, D. & Jones, R.L. (1982). Effect of IAA on growth and soluble cell wall polysaccharides centrifuged from pine hypocotyl sections. Plant Physiol., 69, 323-6.

Tomos, A.D. & Zimmermann, U. (1982). Determination of water relation parameters of individual higher plant cells. In Biophysics of Water. eds. F. Franks, S.F. Mathias pp. 256-61. Chichester: Wiley.

Tomos, A.D., Steudle, E., Zimmermann, U. & Schulze, E-D. (1981). Water relations of leaf epidermal cells of *Tradescantia virginiana*. Plant Physiol., 68, 1135-43.

Tyerman, S.D. & Steudle, E. (1982). Comparison between osmotic and hydrostatic water flows in a higher plant cell: Determination of hydraulic conductivities and reflection coefficients in isolated epidermis of *Tradescantia virginiana*. Aust. J. Plant Physiol., 9, 461-79.

Tyree, M.T. & Jarvis, P.G. (1982). Water in tissues and cells. In Physiological Plant Ecology II. Water Relations and Carbon Assimilation, eds., O.L. Lange, P.S. Nobel, C.B. Osmond & H. Ziegler. Encyl. Plant Physiol., 12B, 35-77.

Tyvand, P. (1982). A hydrodynamic leaf model. Ann. Bot. 50, 799-816.

Van den Ende, J. & Korneef, P. (1960). Gibberellic acid and osmotic pressure. Nature, 186, 327.

Van Volkenburgh, E. & Cleland, R.E. (1980). Proton excretion and cell expansion in bean leaves. Planta, 148, 273-8.

Van Volkenburgh, E. & Cleland, R.E. (1981). Control of light-induced bean leaf expansion: role of osmotic potential, wall yield stress, and hydraulic conductivity. Planta, 153, 572-7.

Van Volkenburgh, E., Schmidt, M.G. & Cleland, R.E. (1985). Limitations of growth in maturing bean leaves. Planta, (in the press).

Von Sachs, J. (1873). Lehrbuch der Botanik, Leipzig (English translation in 1882: Textbook of Botany, 2nd Edition pp. 773-839. Oxford: Clarendon Press).

Weatherley, P.E. (1982). Water uptake and flow in roots. In Physiological Plant Ecology II Water Relations and Carbon Assimilation. eds. O.L. Lange, P.S. Nobel, C.B. Osmond & H. Ziegler. Encyl. Plant Physiol., 12B, 79-134.

Wainwright, S.A., Briggs, W.D., Currey, J.D. & Gosline, J.M. (1976). Mechanical Design in Organisms. London: Edward Arnold.

Westgate, M.E. & Boyer, J.S. (1984). Transpiration- and growth-induced water potentials in maize. Plant Physiol., 74, 882-9.

Yamamoto, R., Makai, K. & Masuda, Y. (1974). Auxin and hydrogen ion actions on light grown pea epicotyl segments. - 3. Effect of auxin and hydrogen ions on stress-relaxation properties. Plant Cell Physiol., 15, 1027-38.

Zimmermann, M.H. & Brown, C.L. (1971). Trees. Structure and Function. Berlin: Springer.

Zimmermann, U. & Steudle, E. (1978). Physical aspects of water relations of plant cells. Adv. Bot. Res., 6, 45-117.

WATER TRANSPORT AS A LIMITING FACTOR IN EXTENSION GROWTH

E. Steudle
Arbeitsgruppe Membranforschung am Institut für Medizin der Kernforschungsanlage Jülich, Postfach 1913, D-5170 Jülich, FRG.

INTRODUCTION

It is generally accepted that under short-term conditions changes in the external water potential mainly affect physical properties of elongating tissues such as the cell turgor (P), the irreversible plastic cell wall extension and the concommitant uptake of water by expanding cells. Even in a humid environment, water transport within and to growing tissues could, in principle, become a rate limiting factor besides others (cell wall extensibility, solute uptake, cell wall and protein synthesis, *etc*.). The question of a rate limitation by water transport is still controversial. Some authors propose that hydraulic resistances in growing tissue may be significant (*e.g.* Molz & Boyer, 1978) whereas others conclude that these resistances are small (*e.g.* Cosgrove & Cleland, 1983). For technical reasons the quantitative evaluation of the hydraulic resistances in growing tissues has been difficult, but with the introduction of the pressure probe technique for higher plant cells (Hüsken *et al*., 1978; see reviews: Zimmermann & Steudle, 1978, 1980; Steudle & Zimmermann, 1984) and with additional techniques for measuring water transport at the tissue level more data are now available of the water relations of growing cells and tissues. This chapter summarises some of the recent aspects of this work.

ROLE OF WATER TRANSPORT DURING STATIONARY GROWTH

The most simple approach to describe extension growth of plant cells and tissues is the Lockhart-equation (Lockhart, 1965) which relates the growth rate to the forces that drive irreversible wall extension. The relation has been worked out in detail for Nitella internodes (Green et al., 1971) where growth responses to changes in cell turgor could be measured directly. For an isolated cell the growth rate (dV/dt; V = cell volume) is given by:

$$\left(\frac{1}{\phi . V} + \frac{1}{Lp . A_o}\right)\frac{dV}{dt} = \pi_c - \pi_{sur} - Y \qquad (1)$$

where ϕ = cell wall extensibility; A_o = cell surface area; Lp = hydraulic conductivity of cell membrane; π_c = osmotic pressure of cell; π_{sur} = osmotic pressure of cell surroundings; Y = yield threshold for plastic extension. Equation (1) states that under stationary conditions, dV/dt is proportional to the total driving force on the right side of the equation. The "growth resistance" is given by the term in brackets on the left side which denotes that there is a mechanical resistance ($1/\phi . V_o$) in series with an hydraulic resistance ($1/Lp . A_o$). It should be noted that under short term conditions V_o and A_o may be taken as constant to a good approximation, because the increases in volume and surface area with time are rather slow. The wall extensibility, ϕ, is a material property of the wall and has the dimensions of a fluidity (inverse viscosity; poise^{-1}). It is assumed that the wall has properties of a "Bingham body" (cf. Scott Blair, 1969). When a certain stress is exceeded the wall behaves like a viscous fluid or paste, i.e. $1/V \, dV/dt = \phi(P-Y)$. The extensibility, ϕ, describes the easiness by which the microfibrils within the wall can slip irreversibly past each other. Properties like stress-hardening or/and processes like creep which are observed in real wall material are not considered.

For plant tissues, the evidence for the validity of equation (1) and for the viscous properties of cell walls is indirect (see reviews by Bradford & Hsiao, 1982, Tyree & Jarvis, 1982). For higher plant tissue the responses of growth to changes in the external ψ are fast (e.g. Acevedo et al., 1971) and, thus, should be mediated by physical rather than by metabolic processes. However, for *Nitella* as well as for higher plants it turned out that ϕ as well as Y were not real constants but depended on environmental factors such as water stress (see Bradford & Hsiao, 1982, and Chapter 1) and shift with time in order to adjust and stabilise growth under various conditions (Green *et al.*, 1977). Furthermore, growth regulators act on the growth rate mainly by increasing the fluidity of the wall. The "acid-growth hypothesis" has been proposed to account for growth regulator effects (Hager *et al.*, 1971; Cleland & Rayle, 1978). According to this hypothesis, growth regulators stimulate H^+-extrusion from growing cells, thus, lowering the wall pH which in turn activates unknown enzymes in the wall responsible for wall loosening.

For tissues, under conditions of stationary growth a relation analogous to equation (1) has been used. The driving force for the growth process may be represented by the water potential of the supplying vessels (ψ_v), the cell osmotic potential (π_c) and the mechanical yield threshold of the cell wall material (Y). On the tissue level, the mechanical and hydraulic resistances will depend on the tissue geometry. For the hydraulic resistance of the tissue, two possible pathways for water have to be considered, *i.e.* the apoplasmic and the cell-to-cell path, whereby the latter comprises the symplasmic as well as the transcellular (vacuolar) pathways (Steudle & Jeschke, 1983). Fig. 1 schematically shows the situation which (under stationary conditions and analogous to equation 1) yields for a growing cell:

$$\left(\frac{1}{\phi \cdot V_o} + \frac{R_{cc} \cdot R_{cw}}{R_{cc} + R_{cw}}\right) \frac{dV}{dt} = \psi_v + \pi_c - Y \qquad (2)$$

The hydraulic resistance of the tissue (second term in the brackets) is divided into two parallel resistors (R_{cc} and R_{cw}) representing the two pathways. R_{cc} and R_{cw} will depend on the hydraulic conductivities (Lp and Lp_{cw}) of both pathways, on the mean cross-sectional areas available for

Fig. 1. Schematical presentation of water flow between vessels and expanding tissue under stationary conditions according to equation (2). Water flow or expansion rate (dV/dt) is limited by a mechanical ($1/\phi . V_o$) and a hydraulic resistance in series. The hydraulic resistance consists of two parallel resistors representing the cell wall and cell-to-cell pathways in the tissue.

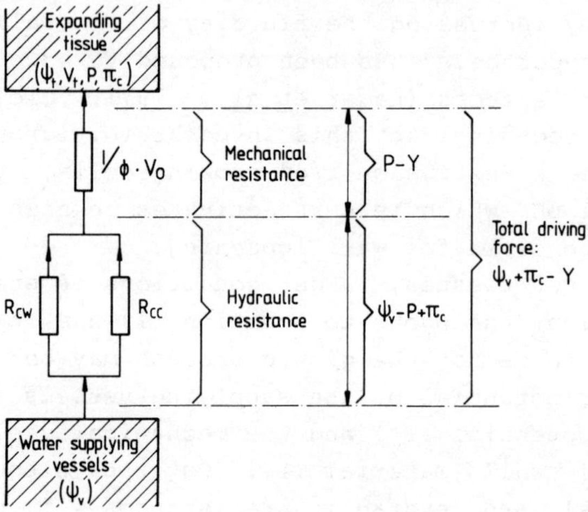

water transport (a_{cc} and a_{cw}, where $a_{cc} >> a_{cw}$) as well as thickness of the tissue which has to be crossed (l):

$$R_{cc} = \frac{2\ l/\Delta x}{a_{cc}\ .Lp} \qquad R_{cw} = \frac{l}{a_{cw}\ .Lp_{cw}} \tag{3}$$

Δx in equation (3) denotes the cell dimension and the factor of two is included since two cell membranes have to be crossed for each of the $l/\Delta x$ cell layers. From Fig. 1 and equation (2) it is evident that either mechanical or hydraulic resistances can become rate limiting. It is also shown that a possible limitation of growth by water transport will depend on the transport mechanism for water in the tissue. For example, if Lp_{cw} is large, the wall component could dominate transport although a_{cw} is generally much smaller than a_{cc}. The latter model is, at present, the most favoured for water transport in higher plant tissue because it is thought that water should move around the cells because the cell Lp is assumed to be rather low. However, measurements of Lp on growing (see below) as well as on other higher plant tissue using the pressure probe (see reviews: Zimmermann & Steudle, 1978, 1980; Steudle et al., 1983; Steudle & Zimmermann, 1984) indicate that Lp could be fairly large so that this picture may be questioned (see below).

As already mentioned, the difficulty in using equations (1) and (2) in growth studies arises from the fact that the mechanical parameters (ϕ, Y) are not constant but are under physical and metabolic control. In principle, variations in Lp and Lp_{cw} are also possible. The equations are, thus, only valid under short-term conditions. There are, however, two other serious difficulties in using these equations. The underlying assumption of a pure viscous behaviour of higher plant cell walls has never been worked out by direct stress/strain measurements on intact tissue and the assumption is perhaps too simple. Secondly, the osmotic pressure component (π_c) is not an independent variable, but is a function of the growth rate (dV/dt). Therefore, π_c will also depend on the hydraulic and mechanical resistances in the system.

There is no doubt that in a growing tissue the walls do show plastic behaviour but the question is whether this is a simple viscous behaviour and irreversible viscoelastic extension. There are attempts to correct for the deviations by introducing additional parameters into equations (1) and (2) which describe time dependent wall softening and hardening processes (cf. Grenetz & List, 1973; see also Tyree & Jarvis, 1982). It may be possible to model growth exactly by increasing the number of parameters. However, it is unlikely that this could prove the underlying assumptions of the process. Furthermore, the physical interpretation of wall extensibility is not clear for objects which consist of quite different tissues and cells. For example, the tensional stress acting on the cell walls

(T) depends not only on cell turgor (P) but also on cell dimensions and cell wall thickness (d). For an elongating cylindrical cell T = P.r/2d (r = cell radius). It has been readily shown that for the elastic properties of cells and tissues geometry factors play an important role (see reviews: Zimmermann & Steudle, 1978, 1980). At a given cell turgor, the elastic modulus (ϵ) increases with increasing cell volume (dimensions) and this is due to a stress-hardening effect, since the wall tension is larger in bigger cells (Steudle *et al*., 1977b; Hüsken *et al*., 1978). Stress-hardening could be also important for the viscous properties of elongating cells. By applying equation (2) to complex tissues we average over cells of different type and size hoping that the mean values still have some significance. The same may be true for averaging the turgor in a tissue. Perhaps, some of the deviations from the linear relationship between growth and turgor are caused by this fact.

In principle, it should be possible to relate the growth rate and turgor (or wall tension) to each other by measuring the parameters in individual cells using the pressure probe, although there are still technical problems which have to be overcome. These problems are related to the determination of growth rates at the cellular level of the order of a few per cent per hour rather than to the probe measurements itself and should be solved. Measurements at the cell level in combination with measurements at the tissue level should allow to judge whether the assumptions made in equations (1) and (2) really hold in the intact tissue.

GROWTH LIMITATION BY SOLUTE TRANSPORT

The second difficulty related to equation (2) is that *via* the "driving force" for growth is a function of the expansion rate (dV/dt), because water uptake dilutes the cell sap. In order to sustain growth, the osmotic pressure of the cells has to be maintained and solutes must be either imported into the cell or generated within the cell (*e.g*. by degradation of storage products) to compensate for the dilution. Stationary π_c requires that the osmolarity of the sap taken up by the growing cells has the same osmolarity as the cell sap, i.e. :

$$\pi_c = RT \frac{J_s'}{J_V'} \tag{4}$$

J_s' (mole.s^{-1}) = solute flow into growing cells; J_V' = dV/dt (cm .s^{-1}) = water flow (expansion rate). J_s' may have active and/or passive components. It is evident that if $RT.J_s'/J_V' < \pi_c$, growth will rapidly cease and if, on the other hand, $RT.J'/J_V' > \pi_c$, the growth rate will increase until the condition of equation (4) is reached. Following the procedure given by Fiscus (1975) and by Dalton et al. (1975) (see also Steudle et al., 1977a) for the stationary uptake of water and solutes into roots, we may combine equations (2) and (4) to obtain:

$$\left(\frac{1}{\phi \cdot V_0} \qquad \frac{R_{cc} \cdot R_{cw}}{R_{cc} + R_{cw}}\right) J_V' = \psi_v + RT \frac{J_s'}{J_V'} - Y \tag{5}$$

In the resulting equation which is quadratic in J_V', the rate of cell expansion would be determined by J_s' provided that ϕ, Y, Lp, Lp_{cw}, ψ_v and the geometry factors are constant. It is possible to split J_s' in equation (5) into active and passive components and to introduce the solute permeability (P_s) and the reflection coefficient (σ_s) for the passive solute uptake analogous to the procedure given by Fiscus (1975, 1977). Equation (5) can be used to demonstrate that in addition to the hydraulic and mechanical resistances considered in Fig. 1, there will be an apparent resistance to growth due to solute uptake. For example, if the growth rate is studied as a function of external water potential (i.e. of ψ_v), the apparent resistance to growth should be given by $d\psi_v/dJ_V'$. In analogy to the root we get from equation (5):

$$\frac{d\psi_V}{dJ_V'} \quad \frac{1}{\phi \cdot V_o} + \frac{R_{cc} \cdot R_{cw}}{R_{cc} + R_{cw}} + RT \frac{J_s'}{(J_V')^2} \tag{6}$$

The additional solute component in the growth resistance on the right side of equation (6) depends on the absolute value of J_s' (or π_c) and rapidly declines with increasing growth rate. This is logical since at high J_s' (π_c) changes in the growth rate have the biggest absolute effect on π_c and the total driving force. J_s' may be the import of inorganic solutes such as K^+ (McNeil, 1976) or of organic solutes delivered from the phloem (Barlow et al., 1980; Meyer & Boyer, 1981). In Avena coleoptiles, carbon dioxide fixation into malate and the parallel uptake of K^+ could provide a sufficient J_s' to drive expansion growth (Haschke & Luttge, 1977). Quantitative measurements of J_s' are necessary to work out the significance of solute flow for growth processes and to relate J_s' to the appropriate J_V'. The pressure probe technique provides a possibility of directly determining solute movements into tissue cells and this technique may be of interest for studying solute relations of growing tissue.

The technique permits the estimation of the time required for the equilibration of solutes between a cell and its surroundings and the evaluation of permeability (P_s) and reflection (σ_s) coefficients in a single experiment (Fig. 2). Upon addition of a permeable solute to the medium the time course of cell turgor changes is biphasic. The first phase is controlled by water relations parameters of the cell (Lp, ϵ; "water phase"), whereas the second phase is governed by P_s ("solute phase"). The ratio between the change in the external osmotic pressure ($\Delta\pi_s^o$) and the change in turgor between original (P_o) and minimum (P_{min}) turgor determines σ_s (Fig. 2; Tyerman & Steudle, 1982; Steudle & Tyerman, 1983; Steudle et al., 1983). This method has been applied to the epidermis of Tradescantia virginiana and to Chara internodes. It has been shown that for the evaluation of P_s, it is not necessary to measure the entire "solute phase" (which could last rather long depending on the absolute value of P_s). It is sufficient to analyse the time

Fig. 2. Pressure/time course of an isolated plant cell after changing the osmotic pressure of the medium by adding a permeable solute (schematical presentation). From the first (water) phase of the biphasic curve Lp can be calculated and from the second (solute) phase the permeability coefficient of the solute (P_s) is obtained. The reflection coefficient (σ_s) is determined from the change in turgor at the minimum (P_o - P_{min}) (redrawn from Steudle et al., 1983).

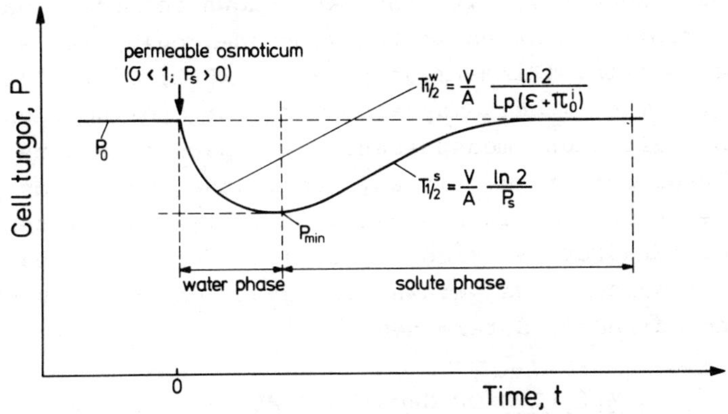

interval around the turgor minimum (Tyerman & Steudle, 1984). This could avoid complications due to irreversible wall extension during measurement. For tissues unstirred layers have to be taken into account when P_s is determined by this method. These effects could be serious or even rate controlling depending on the absolute value of P_s. Nevertheless, the actual values obtained in such experiments would be those which reflect the real situation in a growing tissue.

Maintenance of turgor and adjustment of cell osmotic pressure (π_c) is required for growth to continue under conditions of varying external water potential (e.g. under conditions of water stress). From equation (5) it is evident that a complete adjustment of J_V' under conditions of reduced ψ_v would require an increase in J_s'. Turgor pressure may trigger the uptake of solutes into growing cells. For algal cells it is well established that turgor changes are transformed into changes in active and/or passive membrane transport (see reviews by Zimmermann, 1977, 1978; Bisson & Gutknecht, 1980; Zimmermann & Steudle, 1978, 1980; Steudle & Zimmermann, 1984). The electromechanical model of the cell membrane (Coster & Zimmermann, 1976; Coster *et al.*, 1977; Zimmermann, 1977, 1978) proposes that the changes are mediated by membrane compression and that changes in turgor as well as in the electrical membrane potential could be sensed. Recently, it has been shown for the alga *Valonia* that mobile charges in the membrane could play a role in turgor sensing (Zimmermann *et al.*, 1982; Benz & Zimmermann, 1983). The model may also work for growing higher plant cells, although measurements in expanding tissue have not yet been performed. In any case, the conclusion that the uptake of solutes into growing cells is not related to cell turgor (Bradford & Hsiao, 1982) seems to be too early, since this conclusion was drawn from experiences in which turgor was not directly determined.

LIMITATION OF GROWTH BY WATER TRANSPORT

As already pointed out, the question of a possible limitation of expansion growth by water transport even under conditions of a humid environment is dependent upon the proper transport models for water movement in tissues. The description of water transport in growing tissues in terms of hydraulic and mechanical resistances as shown in Fig. 1 is incomplete because the storage properties of the tissue (cell walls and cells) are not taken into account. They, in fact, may slow down the responses of a growing tissue to changes in the water supply under real conditions and could lead to limitations in growth.

Philip (1958) was the first to present a quantitative theory of water transport through plant tissues. In his model he assumed that water transport in cell walls was negligible and that the transport of water from cell to cell was proportional to the gradient in the water potential. This resulted in a description of water transport in terms of a diffusion type of kinetics, in which the diffusivity of the cell-to-cell path (D_c) was a measure for the rate by which changes in cell volume, turgor or water potential are propagated across the tissue. In Philip's theory the diffusivity of the cell-to-cell path is given by:

$$D_c = \frac{\alpha}{2} Lp.\ \Delta x\ (\epsilon + \pi_c) \sim \frac{1}{T_{1/2}} \quad (7)$$

where α = shape factor; Lp = hydraulic conductivity of the cell membrane; Δx = cell diameter in the direction of the transport; ϵ = elastic modulus of cell and π_c = osmotic pressure of cell. It can easily be shown that the diffusivity would be inversely proportional to the half-time of an individual cell for water exchange ($T_{1/2}$) which can be, therefore, directly determined using the pressure probe.

Molz & Ikenberry (1974) extended Philip's theory and allowed for the water flow in the cell wall as well as the classical cell-to-cell pathway. They assumed that during transport in the parallel pathways the cells were in local equilibrium with the adjacent wall space. This condition seems to be fullfilled to a very good approximation for higher plant cells since the $T_{1/2}$ values measured so far with the aid of the pressure probe show (cf. reviews by Zimmermann & Steudle, 1978, 1980; Steudle & Zimmermann, 1984) that $T_{1/2}$ values are usually of the order of a few seconds. For growing tissues of pea and soybean values of $T_{1/2}$ = 0.3 to 1.2 s and $T_{1/2}$ = 0.3 to 15.1 s, respectively, have been found (Cosgrove & Steudle, 1981; Steudle & Boyer, 1985). The diffusivity of a tissue (D) comprising both pathways will be given by (Steudle & Boyer, 1984):

$$D_t = \frac{\Delta x\ (\sigma_{cw} .Lp_{cw} .a_{cw} \quad Lp.a_{cc} .\Delta x/2)}{C + C_{cc}} \quad (8)$$

where σ_{cw} is the reflection coefficient of the wall path for the wall solutes. C_{cw} and C_{cc} are the storage capacities of the cell wall and cell-to-cell pathway, respectively. Fig. 3

shows an electrical analog diagram of the tissue in terms of the Molz and Ikenberry theory. It is clear from the figure that for sufficiently small values of C_{cw} and C_{cc} the model reduces to two parallel resistors (Fig. 1). Thus, the Molz and Ikenberry theory extends and completes the description of tissue water transport. It should be noted that in our description the symplasmic pathway was incorporated into the cell-to-cell path because (for experimental reasons) the hydraulic conductance of the plasmodemata within the cell membrane is very difficult to separate from that of the overall hydraulic conductivity of the cell. In theory, however, the symplasmic path may be incorporated (Molz et al., 1979).

Fig. 3. Network analog of the cell-to-cell/cell wall transport system. C_{cc} and C_{cw} are the water storage capacities per cell of the parallel pathways (see equation (8)). r_{cw} and r_{cc} can be identified with $r_{cw} = \Delta x/a_{cw} . Lp_{cw}$ and $r_{cc} = 2/Lp.a_{cc}$ (compare equation (3)), respectively, whereas the resistance r denotes the hydraulic resistance between cell and adjacent wall space (redrawn from Molz & Ikenberry, 1974).

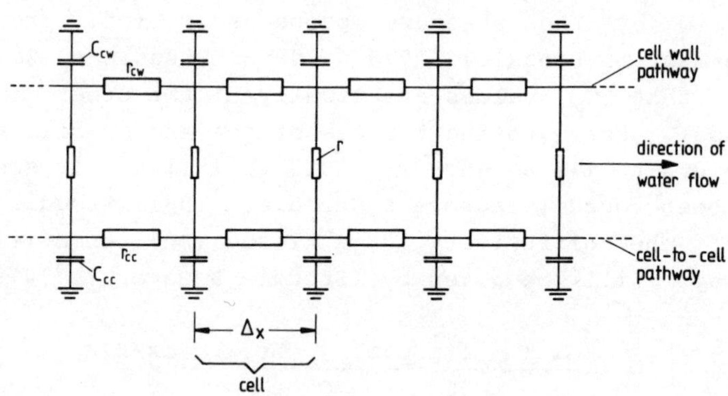

Molz and Boyer (1978) have applied the theory of Molz and Ikenberry to growing tissue and have analysed the water supply to growing tissue in the cylindrical hypocotyl of soybean seedlings. They assumed literature data for the hydraulic conductivities, cross-sectional areas and storage capacities and arrived at the conclusion that under conditions of steady growth considerable gradients in the water potential may exist within the tissue and that, therefore, a limitation of growth by water transport cannot be excluded. The calculated growth-induced water potential gradients were as large as 0.3 MPa between epidermis and supplying vessels, i.e. over a distance of only 0.5 mm. The gradients measured with the isopiestic psychrometer were in fairly good agreement with the calculated values (Molz & Boyer, 1978). Furthermore, gradients were also found experimentally between mature and elongating regions of the seedling (Cavalieri & Boyer, 1982). Auxin caused the gradients to drop which has been interpreted by an increase of Lp caused by auxin (Boyer & Wu, 1978).

Cosgrove & Steudle (1981) determined water relations in growing pea epicotyl at the cell and tissue level and found a D_c value of $3.2 \times 10^{-10}\ m^2.s^{-1}$ from $T_{1/2}$ values. D_t measured from shrinking and swelling experiments of pealed epicotyl segments was $6.6 \times 10^{-10}\ m^2.s^{-1}$. From these data the authors concluded that water transport during shrinking or swelling and during growth was from cell to cell to a considerable extent. They also concluded that there could be a limitation of growth by water transport, at least to some extent. According to the Molz and Boyer theory the gradients across the cortex of the pea epicotyl were 0.02 and 0.1 MPa for growth rates of 1% h^{-1} and 5% h^{-1}, respectively. Cosgrove and Steudle (1981) point out that, although Lp values of the cells were large ($Lp = 0.4$ to $9 \times 10^{-6}\ m.s^{-1}.MPa^{-1}$) and $T_{1/2}$ short ($T_{1/2} = 0.3$ to 1.2 s), the gradients may be maintained because water has to move across a rather large number of cell layers. Cosgrove & Cleland (1983) in repeating most of the experiments of Cosgrove & Steudle (1981) found similar data for the pea epicotyl. In addition, they also performed measurements on the effect of auxin on the cell Lp but did not see an effect of IAA on Lp.

They concluded that gradients within the tissue should be very small if not negligible. However, they give no reasons for this conclusion, e.g. on the basis of a different transport model.

Steudle & Boyer (1985) have tried to decide the question of a possible limitation of growth by water transport by using different approaches. Using growing hypocotyls of soybean they measured Lp at the cell level using the pressure probe, the tissue diffusivity (D_t) by sorption experiments and the radial hydraulic conductance of the cortical tissue (Lp_r) by a new pressure-perfusion technique. The latter technique (Fig. 4) permits the measurement of artificially imposed constant water flows either from the xylem radially to the outside (in to out experiments) or from the outside into the xylem (out to in experiments). Simultaneously the determination of the driving hydrostatic pressure gradient is possible. Pressure/volume curves obtained by this procedure were linear for sufficiently low water flows and revealed the same Lp independent of the direction of flow. Lp_r values strongly depended upon the existence of a cuticle. With intact segments still having a cuticle Lp (per cm^2 of outer surface) was Lp_r = (0.1-3.7) x 10^{-6} $m.s^{-1}.MPa^{-1}$ and with the cuticle abraded Lp_r = (2-20) x 10^{-6} $m.s^{-1}.MPa^{-1}$. For both growing and mature regions similar Lp values were obtained. Flooding of the tissue occurred when the segments were perfused for several hours. However, experiments with abraded segments which had been infiltrated in the vacuum showed that flooding did not change Lp significantly. The results with the abraded segments show that Lp was of the same magnitude or even larger than the single cell Lp (Lp = 0.2 to 10 x 10^{-6} $m.s^{-1}.MPa^{-1}$). This result is incompatible with a significant cell-to-cell transport in the cortex and suggests a dominating apoplasmic transport under the conditions of hydrostatic perfusion. At first sight, the result not only contradicts probe measurements on the pea epicotyl (see above) and on roots (Steudle & Jeschke, 1983), where a considerable contribution of the

Fig. 4. Pressure-perfusion technique for measuring radial hydraulic conductance of hypocotyl segments of soybean seedlings. A) In to out experiment: Constant water flows are pumped from a cut surface into the xylem and across the cortex. B) Out to in experiment: Flows are forced from outside into the xylem and exude into a collector. Water flowing across the cortex in both directions is collected in calibrated tubes. The stationary hydraulic pressures driving the flows are measured with the aid of pressure transducers within the pressure chambers (taken from Steudle & Boyer, 1985).

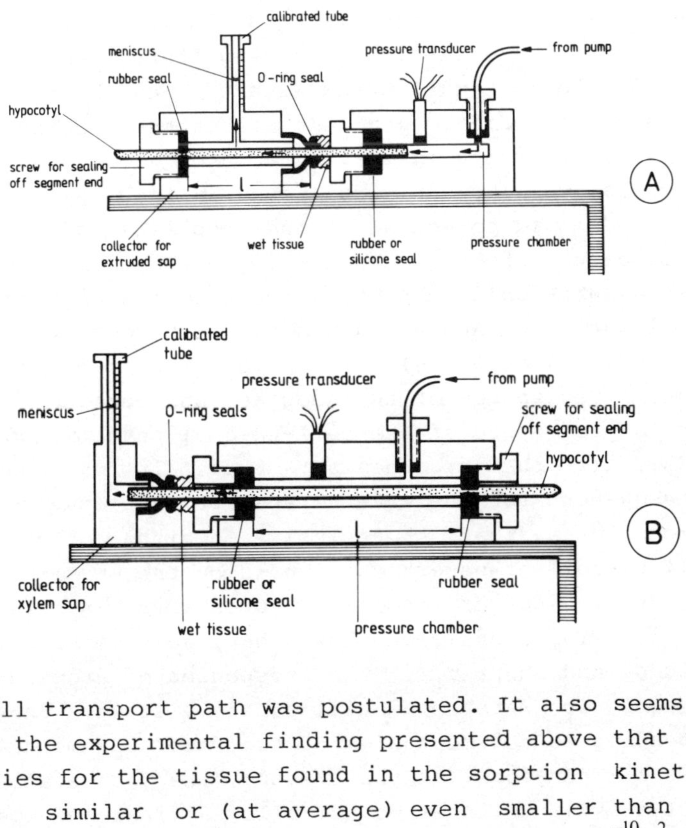

cell-to-cell transport path was postulated. It also seems to contradict the experimental finding presented above that the diffusivities for the tissue found in the sorption kinetics (D_t) were similar or (at average) even smaller than D calculated from probe data (D_t = 0.2 to 0.8 x $10^{-10} m^2.s^{-1}$; D_c = 0.1 to 5.5 x $10^{-10} m^2.s^{-1}$). If Lp during sorption and growth would be of the same order or even larger than Lp of the cells, one would expect that the tissue reacts with a half-time of a few seconds instead of a few minutes.

A tentative explanation for the phenomenon is that during hydrostatic perfusion the wall path is dominating and has a very high hydraulic conductivity (Lp_{cw}). For osmotic experiments as well as for growth the cell-to-cell path dominates. The driving force in the apoplast under these conditions should be an osmotic gradient rather than a gradient in the matric potential, since the wall space is very close to water saturation, and in the perfusion fluid significant amounts of solutes were found (Steudle & Boyer, 1985). The osmotic gradient should be established by the fact that the cells (which have different water potentials) are in local equilibrium with the adjacent wall space. However, an osmotic gradient within the apoplast should not be very effective in driving a water flow across the wall space since the driving force will be given by $\sigma_{cw} \cdot \Delta\pi_{cw}$ (π_{cw} = osmotic potential in the apoplast) and the reflection coefficient of the cell wall (σ_{cw}) should be very low. If this assumption is correct, the theories of water transport in tissues (see above) should be extended by incorporating σ_{cw} (see equation 8). This would result in an apparently higher resistance of the apoplast and could be the reason for a preference of the cell-to-cell path (Steudle & Boyer, 1985). If the hypothesis is true, it may have general consequences for the water transport in tissues, because it demonstrates that the relative magnitude of cell wall vs. cell-to-cell transport can depend on the physical nature of the driving forces (osmotic, hydrostatic) involved.

This example shows that, despite high Lp and Lp_{cw} values and short $T_{1/2}$, the responses of growing tissue to changes in water potential could be slower than expected depending on the mechanism of the water movement within the tissue. Therefore, at the present, the data available at the cell and tissue level do not allow a final judgement about a limitation of growth by water transport. To exclude water transport as a limiting intrinsic factor within a growing tissue would be premature. We do not have the right models for the water relations at the tissue and organ level.

CONCLUSIONS

1. Experimental and theoretical evidence shows that the Lockhart-equation (equations 1 and 2) normally used to describe extension growth of plant cells and tissues should be checked for its validity in the analysis of growth processes. Especially, the underlying assumption of ideal viscous properties of the cell wall of turgid cells should be checked by measurements at the cell level and compared with the relations found on the tissue or organ level.

2. The significance of solute transport in expanding cells should be also checked quantitatively. The pressure probe technique offers some possibilities in doing this.

3. Water relations data collected so far for growing tissues at the cell level suggest that Lp is large and $T_{1/2}$ short which may point to a necessity in these tissue for having low hydraulic resistances. The hydraulic conductance of the cell wall path could be very high.

4. Despite of high Lp and Lp_{cw} values and short $T_{1/2}$, the equilibration of water potential gradients during growth could be restricted. There is evidence that during growth the cell-to-cell transport dominates and, thus, the hydraulic resistances within a growing tissue could become significant depending on the absolute value of Lp and on the number of cell layers which have to be crossed.

5. It is necessary to further develop models for the water transport in growing tissues and to determine the relevant parameters (Lp, Lp_w, $T_{1/2}$, D_c, D_t etc.) using the new techniques in order to answer the question of the relative importance of the mechnical vs hydraulic resistances and of the significance of the cell wall vs cell-to-cell hydraulic resistance.

ACKNOWLEDGEMENTS

I thank Dr M.E. Westgate for carefully reading the manuscript. This work was supported by a grant from the Deutsche Forschungsgemeinschaft Ste 319/1.

REFERENCES

Acevedo, E., Hsiao, T.C. & Henderson, D.W. (1971). Immediate and subsequent growth responses of maize leaves to changes in water status. Plant Physiol., 48, 631-636.

Barlow, E.W.R., Munns, R.E. & Brady, C.J. (1980). Drought responses to apical meristems. In Adaptation of plants to water and high temperature stress, eds. N.C. Turner, P.J. Kramer, pp. 191-205. New York: Wiley and Sons.

Benz, R. & Zimmermann, U. (1983). Evidence for the presence of mobile charges in the cell membrane of *Valonia utricularis*. Biophys. J. 43, 13-26.

Bisson, M.A. & Gutknecht, J. (1980). Osmotic regulation in algae. In Plant membrane transport: current conceptual issues, eds. W.J. Lucas, R.M. Spanswick & J. Dainty, pp. 131-146. Amsterdam: Elsevier/North-Holland.

Boyer, J.S. & Wu, G. (1978). Auxin increases the hydraulic conductivity of auxin-sensitive hypocotyl tissue. Planta, 139, 227-237.

Bradford, K.J. & Hsiao, T.C. (1982). Physiological repsonses to moderate water stress. In Encycl. Plant Physiol., New Series, Vol. 12B, eds. O.L. Lange, P.S. Nobel, C.B. Osmond & H. Ziegler, pp. 263-324. Berlin. Heidelberg, New York: Springer-Verlag.

Cavalieri, A.J. & Boyer, J.S. (1982). Water potentials induced by growth in soybean hypocotyls. Plant Physiol., 69, 492-496.

Cleland, R.E. & Rayle, D.L. (1978). Auxin, H^+-excretion and cell elongation. Bot. Mag. Tokyo Spec. Issue, 1, 125-139.

Cosgrove, D.J. & Cleland, R.E. (1983). Osmotic properties of pea internodes in relation to growth and auxin action. Plant Physiol., 72, 332-338.

Cosgrove, D.J. & Steudle, E. (1981). Water relations of growing pea epicotyl segments. Planta, 153, 343-350.

Coster, H.G.L. & Zimmermann, U. (1976). Transduction of turgor pressure by cell membrane compression. Z. Naturforsch., 31c, 461-463.

Coster, H.G.L., Steudle, E. & Zimmermann, U. (1977). Turgor pressure sensing mechanism. Plant Physiol., 58, 636-643.

Dalton, F.N., Raats, P.A.C. & Gardner, W.R. (1975). Simultaneous uptake of water and solutes by plant roots. Agron. J., 67, 334-339.

Fiscus, E.L. (1975). The interaction between osmotic and pressure induced water flow in plant roots. Plant Physiol., 55, 917-922.

Fiscus, E.L. (1977). Determination of hydraulic and osmotic

properties of soybean root systems. Plant Physiol., 59, 1013-1020.

Green, P.B., Erickson, K. & Buggy, J. (1971). Metabolic and physical control of cell elongation rate. In vivo studies in Nitella. Plant Physiol., 47, 423-430.

Green, P.B., Bauer, K. & Cummins, W.R. (1977). Biophysical model for plant cell growth: auxin effects. In Water relations in membrane transport in plants and animals, eds. A.M. Jungreis, T.K. Hodges, A. Kleinzeller, S.G. Schultz, pp. 30-45. London, New York: Academic Press.

Grenetz, P.S. & List, A. (1973). A model for predicting growth responses of plants to changes in external water potential: Zea mays, primary root. J. Theor. Biol., 39, 29-45.

Hager, A., Menzel, K. & Krauss, A. (1971). Versuche und Hypothese zur Primarwirkung des Auxins beim Streckungswachstum. Planta, 100, 47-75.

Haschke, H. -P. & Luttge, U. (1977). Action of auxin on CO_2 dark fixation in Avena coleoptile segments as related to elongation growth. Plant Sci. Lett., 8, 53-58.

Husken, D., Steudle, E. & Zimmermann, U. (1978). Pressure probe technique for measuring water relations of cells in higher plants. Plant Physiol., 61, 158-163.

Lockhart, J.A. (1965). An analysis of irreversible plant cell elongation. J. Theor. Biol. 8, 264-276.

McNeil, D.L. (1976). The basis of osmotic pressure during expansion growth in Helianthus annuus hypocotyls. Aust. J. Plant Physiol., 5, 589-593.

Meyer, R.F. & Boyer, J.S. (1972). Sensitivity of cell division and cell elongation to low water potentials in soybean hypocotyls. Planta, 108, 77-87.

Molz, F.J. & Ikenberry, E. (1974). Water transport through plant cells and cell walls: Theoretical development. Soil Sci. Soc. Am. Proc., 38, 699-704.

Molz, f.J. & Boyer, J.S. (1978). Growth-induced water potentials in plant cells and tissues. Plant Physiol., 62, 423-429.

Molz, F.J., Kerns, D.V., Peterson, C.M. & Dane, J.H. (1979). A circuit analogy model for studying quantitative water relations of plant tissues. Plant Physiol., 64, 712-716.

Philip, J.R. (1958). Osmosis and diffusion in tissues: Half-time and internal gradients. Plant Physiol., 33, 275-278.

Scott Blair, G.W. (1969). Elementary rheology. London, New York: Academic Press.

Steudle, E., Zimmermann, U. & Lelkes, P.I. (1977a). Volume and pressure effects on the potassium fluxes of

Valonia utricularis. In Transmembrane ionic exchange in plants, eds. M. Thellier, A. Monnier, M. Demarty & J. Dainty, pp. 123-137. Paris: CNRS.

Steudle, E., Zimmermann, U. & Lüttge, U. (1977b). Effect of turgor pressure and cell size on the wall elasticity of plant cells. Plant Physiol., 59, 285-289.

Steudle, E. & Jeschke, W.D. (1983). Water transport in barley roots. Planta, 158, 237-248.

Steudle, E. & Tyerman, S.D. (1983). Determination of permeability coefficients, reflection coefficients, and hydraulic conductivity of Chara corallina using the pressure probe: Effects of solute concentrations. J. Membrane Biol., 75, 85-96.

Steudle, E., Tyerman, S.D. & Wendler, S. (1983). Water relations of plant cells. In Effects of stress on photosynthesis, eds. R. Marcelle, H. Clijsters, M. van Poucke, pp. 95-109. The Hague, Boston, London: Martinus Nijhoff, Dr W. Junk.

Steudle, E. & Boyer, J.S. (1985). Hydraulic resistance to radial water flow in growing hypocotyl of soybean measured by a new pressure-perfusion technique. Planta, in press.

Steudle, E. & Zimmermann, U (1984). Water relations of plant cells: Further development of the pressure probe and of techniques for measuring pressure-dependent transport. In Membrane transport in plants, eds. W.J. Cram, K. Janáček, R. Rybora & K. Sigler, pp. 73-82. Prague: Academia.

Tyerman, S.D. & Steudle, E. (1982). Comparison between osmotic and hydrostatic water flows in a higher plant cell: Determination of hydraulic conductivities and reflection coefficients in isolated epidermis of Tradescantia virginiana. Aust. J. Plant Physiol., 9, 461-480.

Tyerman, S.D. & Steudle, E. (1984). Determination of solute permeability in Chara internodes by a turgor minimum technique. Plant Physiol., 74, 464-468.

Tyree, M.T. & Jarvis, P.G. (1982). Water in tissues and cells. In Encycl. Plant Physiol., New Series, Vol. 12B, eds. O.L.Lange, P.S. Nobel, C.B. Osmond & H. Ziegler, pp. 35-77. Berlin, Heidelberg, New York: Springer-Verlag.

Zimmermann, U. & Steudle, E. (1975). The hydraulic conductivity and volumetric elastic modulus of cells and isolated cell walls of Nitella and Chara spp.: Pressure and volume effects. Aust. J. Plant Physiol., 2, 1-12.

Zimmermann, U. (1977). Cell turgor regulation and pressure-mediated transport processes. In Proc. 31st Symp. Soc. Exp. Biol.: Integration of activity in higher plants, ed. D. Jennings, pp. 117-154. Cambridge:

Cambridge University Press.

Zimmermann, U. (1978). Physics of turgor- and osmoregulation. Annu. Rev. Plant Physiol., 29, 121-148.

Zimmermann, U. & Steudle, E. (1978). Physical aspects of water relations of plant cells. Adv. Bot. Res. 6, 45-117.

Zimmermann, U. & Steudle, E. (1980). Fundamental water relations parameters. In Plant membrane transport: Current conceptual issues, eds. W.J. Lucas, R.M. Spanswick & J. Dainty, pp. 113-127. Amsterdam: Elsevier, North Holland Biomedical Press.

Zimmermann, U. Büchner, K. -H. & Benz, R. (1982). Transport properties of mobile charges in algal membranes: Influence of pH and turgor pressure. J. Membrane Biol., 67, 183-197.

GROWTH SUBSTANCES AND LEAF GROWTH

J. Digby and R.D. Firn
Department of Biology, University of York,
York, YO1 5DD, United Kingdom.

INTRODUCTION

It is just over 50 years since Gregory (1926) proposed that leaf growth might be limited by the lack of a specific regulating chemical. In an attempt to explain why leaf expansion in etiolated plants was minimal, he proposed that leaves in darkness did not expand because they could not produce a necessary chemical unless exposed to light. Thus, the concept of a non-nutritional factor controlling leaf expansion entered the thinking of physiologists. A few years later the first plant growth substance, auxin, was discovered and interestingly it was first shown to have a profound effect on growth of the coleoptile, an organ which is regarded as being a modified leaf. When applied to normal leaves, however, auxin was found to have little effect on overall expansion but it sometimes produced somewhat distorted leaves by promoting the elongation of the veins. Such observations eventually led to the proposal that another regulator must exist in leaves, this "phyllocaline" being responsible for controlling mesophyll growth whilst auxin controlled vein growth (Went, 1951). By evoking the control of leaf growth by two regulators, it was felt that an explanation of leaf form and leaf size could be offered. The idea of multi-regulator control of leaf growth was thereby established. It is evident, therefore, that the hypothesis that leaf growth is regulated by some specific substance or substances was originally based on fairly well defined ideas. However, these ideas did not result in the isolation of any specific leaf growth regulator and there is no compelling evidence that leaves contain any such specific leaf growth regulators or "phyllocalines". Indeed studies of the involvement of plant growth substances in the control of leaf development soon became a minor part of plant growth substance research because many other organs showed much more dramatic responses to applied regulators. As new groups of plant growth substance were discovered in the 1950's and 1960's their effects on leaf growth were usually noted but intact leaves were not the organs of choice for studying any type of regulator. However, interest in the role of plant growth substances in leaves was sustained by the fact that each of the regulators could evoke some response, even if minor, in leaves and the fact that leaves were shown to contain all types of plant growth substance.

THE SIGNIFICANCE OF THE PRESENCE OF GROWTH SUBSTANCES IN LEAVES

There is now satisfactory evidence that each of the major groups of plant growth substance (auxins, gibberellins, cytokinins, ethene and abscisic acid) are present in leaves (Goodwin, 1978; Goodwin & Erwee, 1983). There are, however, a number of possible reasons why leaves might contain plant growth substances:-

(i) they could be regulating leaf expansion

(ii) they could be regulating some aspect of leaf functioning

(iii) they could be synthesised in the leaf but destined to serve as true hormones, acting elsewhere in the plant

(iv) they could be present in the leaf simply as a result of being present in a transport path leading to the leaf and they may not be purposefully directed to the leaf or serve any purpose in the leaf

(v) they may be residual, having served some previous role.

It is apparent that the mere presence of any growth

substance is of little significance in itself. Indeed, the failure to find a particular plant growth substance in a leaf would be of more significance but analytical techniques have become so sensitive over recent years that such conclusive results become ever less likely! Consequently, having found a particular regulator in a leaf further experiments must be conducted to eliminate some of the alternatives listed above. It might seem most sensible to concentrate simply on (i), designing a series of experiments to show for instance that the appropriate amount of the growth substance is in the correct part of the leaf at the time expected (see later section on Correlative Studies). There is, however, some value in attempting to eliminate the alternatives (ii) to (v) as a complementary approach. All too often researchers studying the involvement of one plant growth substance in one physiological process ignore the fact the same substance may have another role in the same organ. The possibility that some of these postulated roles may conflict is not always given adequate consideration. As an example that such alternative roles are realistic consider the case of abscisic acid (ABA) in leaves. ABA is usually termed a growth inhibitor hence it might be thought that it could play some role in regulating leaf expansion. However, the ABA concentration in young rapidly expanding leaves is actually higher than in old leaves (Zeevaart, 1977) and whilst ABA can inhibit leaf growth it only does so at high concentrations (van Volkenburgh & Davies, 1983). In contrast, the fact that ABA levels increase quite rapidly in leaves of plants subjected to water stress and that applied ABA can quickly close stomata is consistent with the view that the major role for ABA in leaves is in modulating leaf water status (Wright, 1978). The known effects of water stress on leaf expansion seem adequately explained by a direct effect of water potential changes on cell extension, rather than an effect mediated via ABA as an inhibitor of cell expansion (Hsiao, 1973). It has also been argued recently that leaf ABA might be the source of ABA found in some developing seeds (Brenner *et al.*, 1983). It can be seen that alternatives (ii), (iii) and (v) seem more satisfactory explanations than (i) for the presence of ABA in leaves.

THE SIGNIFICANCE OF GROWTH RESPONSES IN LEAVES FOLLOWING GROWTH SUBSTANCE APPLICATION

The relatively minor effect that plant growth substances, applied at physiological concentrations, have on the growth of normal leaves might reasonably be regarded as evidence that leaf growth is usually not limited by the supply of any of the substances. However, such straightforward deductions are usually challenged when there are other reasons to believe that a particular regulator might serve some role. The fact that applied auxin usually had a relatively minor effect on shoot extension when applied to intact plants (Thimann, 1977), yet was generally believed to be involved in regulating shoot extension, had set a precedent for workers not being discomforted by inconsistencies of this type. The lack of an effect when a growth substance is applied directly to leaves could quite reasonably be explained by any of the following alternatives:-

1. poor uptake of the substance into the correct compartment of the leaf
2. rapid metabolism of the applied substance
3. rapid movement of the applied substance out of the leaf
4. the presence of optimal amounts of the endogenous substance at the time of application of the exogenous chemical

These are simply alternatives for consideration and possible investigation and one should avoid the trap of accepting, on slender evidence, any of these alternatives rather than rejecting the hypothesis that a substance does control leaf growth. The first alternative is very difficult to dismiss because the correct compartment, be it at the cell or subcellular level, can rarely be identified let alone analysed. The second and third alternatives should be easily studied now that representatives of all the major groups of plant growth substance incorporating a radioactive isotope are readily available. The fourth alternative is possibly best investigated in plants deficient in a particular regulator if such plants can be obtained. Such studies will now be considered.

LEAF DEVELOPMENT IN PLANTS WITH REDUCED GROWTH SUBSTANCE LEVELS

Studies on Grasses

Shortly after the discovery of the gibberellins, it was shown that some dwarf mutants of maize would develop normal stature if supplied with exogenous gibberellin. These mutants have now been well characterised and it has been shown that the mutants have specific blockages of the gibberellin biosynthetic pathway (Phinney & Spray, 1983). Less well characterised dwarf mutants of rice, wheat and ryegrass have also been reported (Goodwin, 1978). In the case of the dwarf maize mutants it is apparent that the leaf length, leaf sheath length and overall plant height are reduced and gibberellin application restores the elongation of all the organs influenced by the mutations. A number of studies with chemical inhibitors of gibberellin biosynthesis in plants have also demonstrated a reduction in leaf blade growth, a reduction usually reversed by gibberellin. Such studies support the work on mutants but the specificity of the inhibitors is rarely absolute, hence the deductions are less secure. However, in all these studies on gibberellin deficient mutants or on chemically treated plants, the fact that applied gibberellin restores leaf elongation to normal does not prove that gibberellin restored leaf growth by direct action on the leaf. It is known that gibberellins are involved in a number of physiological responses in the plant and it is possible that some of these responses may influence leaf growth (see Concluding Remarks). Consequently, reduced leaf growth in plants with reduced gibberellin levels could be a secondary effect and the restoration of leaf growth could be the result of gibberellin action outside the leaf. Some studies on the kinetics of the effect of gibberellins on leaf growth in the maize mutants are needed; one would expect a rapid restoration of growth rate if gibberellin treatment acted directly on leaf expansion. An effect of gibberellin application on the elongation rate of an isolated leaf would be even more conclusive.

Studies on Dicotyledons

It must be borne in mind that monocotyledon and dicotyledon leaf growth is so different that one should not necessarily expect comparable results from the two groups of plants. Interestingly there is indeed much less evidence for a role for gibberellins in controlling leaf growth in dicotyledons. For instance, mutants of pea which are dwarf in habit and some of which have low gibberellin levels (Potts & Reid, 1983) seem not to be noted for consistently abnormal leaf growth, although a full study is awaited. Dicotyledons treated with growth retardants (thereby possibly containing reduced gibberellin levels) also show no consistent effect on leaf size, increases and decreases in lamina area being reported. In some cases a clear reduction in leaf growth after growth retardation treatment is evident and in a number of studies this reduction is reversed by gibberellin treatment but evidence for this being a direct gibberellin effect on the leaf is still required.

Turning our attention now to the possible role of ethene in leaf growth, some studies on the tomato mutant *diageotropica*, which is altered in some respect of its ethene physiology, might provide useful information about the role of this plant growth substance in leaf growth or development. The mutant has dark green leaves which show signs of hyponasty (the leaf edges curled up somewhat). Zobel (1973) showed that leaf growth could be restored to normal if the mutant was grown in an atmosphere containing a low concentration of ethene. The application of IAA to the mutant could also restore the mutant to form normal leaves, possibly as a result of IAA-induced ethene production. In the case of ethene supplementation or IAA addition, only the growth of new leaves was restored to normal and the hyponasty of the existing leaves was not fully overcome. Unfortunately the exact nature of the physiological abnormality of diageotropica is still somewhat uncertain (Jackson, 1979; Bradford & Yang, 1980), hence firm conclusions are not yet possible about the role of ethene in the development and expansion of normal plants.

Another mutant of tomato, flacca, has an abnormality of its ABA physiology but it apparently shows no dramatic leaf abnormalities (Goodwin, 1978). This observation supports the view that ABA has no significant direct role in controlling leaf growth.

CORRELATIVE STUDIES ON GROWTH SUBSTANCES IN LEAVES

The commonest approach adopted when attempting to assign a role to a particular plant growth substance in a developmental process is to study the relationship between the concentration of the substance and the rate at which the process is proceeding. Whilst this approach seems straightforward, it is in practice full of uncertainties. There is even a debate about whether the concentration of growth substances is the limiting factor in any developmental process, the opposing view being that sensitivity changes to a constant amount of the substance are more important (Cleland, 1983; Trewavas, 1983). However, if one puts aside these doubts and accepts that growth substance concentration is of importance, it can be argued that simply extracting leaves of different ages or sizes and measuring the concentration of all the growth substances in an attempt to find a correlation of some type is a misconceived approach. The reason for this pessimistic view is that leaf development and leaf growth are complex processes. We shall firstly digress slightly to consider leaf development and then we shall return to consider the implications of this information to studies of growth substances in leaves.

Leaf Development and Expansion

In considering leaf expansion as a physiological process it is necessary to think about the stages of anatomical development which a leaf passes through during its growth from primordium to maturity. This is often largely ignored in physiological studies on growth, with the only parameter measured being an increase in area of the lamina or more commonly an increase in length of the whole leaf. Such aspects as rates of cell division and expansion, cell differentiation and growth in thickness of the leaf are often completely ignored. It must also be noted that monocotyledon and dicotyledon leaves are different in several respects with regards their development. In both cases the lamina are developed from the leaf primordium, however, the older leaves in the Graminae are composed of a leaf blade and below it a leaf sheath, and it has been suggested that the primordium itself produces only the leaf blade, with the leaf sheath being produced later by the activity of a separate intercalary meristem lower down the apex (Sherman, 1942; Esau, 1953). It would seem certain then that the leaf sheath of a grass leaf is different in nature from the leaf blade in a dicotyledon, and therefore one would not expect information from the two systems to be readily comparable. Some workers have taken a more extreme view when considering comparability between the two types of leaf and have suggested that the monocotyledonous leaf has no counterpart of the lamina of the dicotyledonous leaf (Arber, 1950). It is clear that lamina development in the Graminae proceeds in a linear manner, with cells being cut off by a basal meristem (Fig. 1). This means that leaf sections taken successively above the meristem will be of increasing developmental age. The zone of cell division is very small and the zone of elongation is also small; *e.g.* cell elongation in maize is restricted to a small zone near the ligule - a zone only 2cm in length in the leaf illustrated in Fig. 1. In very young leaves the situation is further complicated by the fact that the rate of cell production changes with time. Boffey *et al.* (1980) have

Fig. 1. Elongation is restricted to a small zone of an elongating grass leaf. In a maize leaf over 50cm in length, the elongating zone is near the base of the leaf and is only about 5cm long. From Michelena & Boyer (1982).

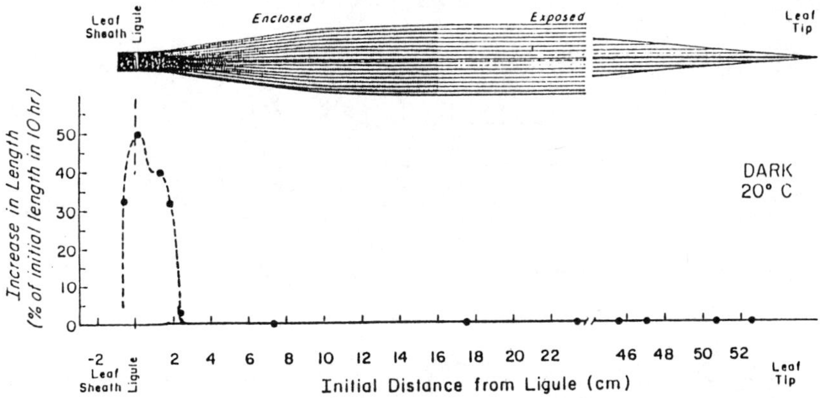

Fig. 2. Growth in a light-grown wheat leaf 4, 5, 6 and 7 days after sowing to illustrate the relative positions and cell sizes of cells A, B, C and D cut off from the basal meristem at intervals of 1 day. Cells B and C move through the elongation zone more rapidly than cell A due to increased activity of the basal meristem with time. From Boffey et al. (1980).

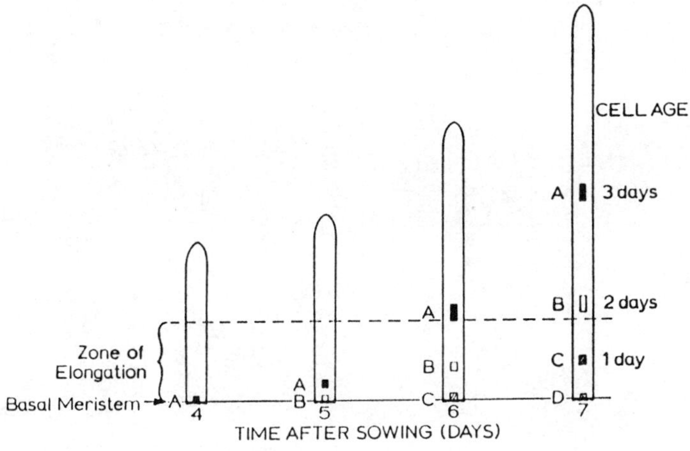

Fig. 3. Diagram illustrating the initial formation of the lamina in a Xanthium leaf. Ab L, abaxial layer; Ad L, adaxial layer; LP, lower protoderm (epidermis); ML, middle layer; UP, upper protoderm (epidermis). From Maksymowych (1973).

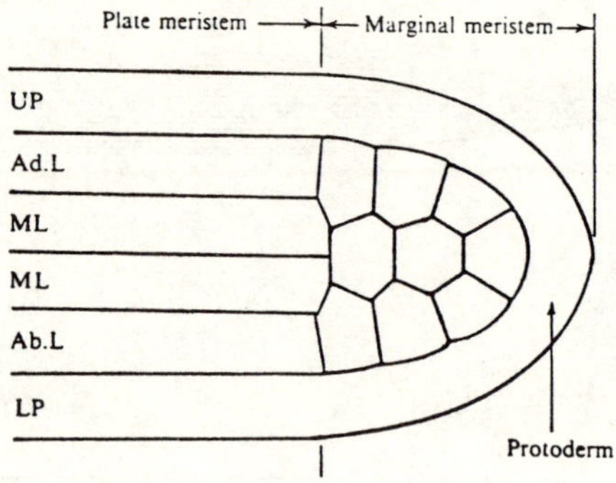

Fig. 4. Stages of lamina development at the tip, middle and base of a Xanthium leaf with increasing developmental age, which is expressed as leaf plastochron index, LPI. From Maksymowych (1973).

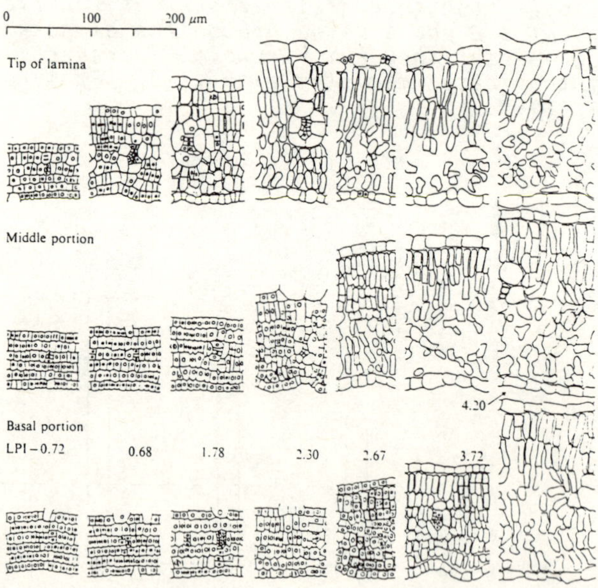

shown that although the zone of elongation remains of constant width with time, the speed at which a cell passes through the elongation zone decreases with time, due to an increase in cell division rate in the meristem (Fig. 2). Development of dicotyledonous leaves is much more complex. Maksymowych (1973) made an exhaustive study of Xanthium leaf development which serves as a useful model (Table 1). Lamina formation begins at a very early stage (LPI-5) and cells are produced by division of the marginal meristem which produces a six layered structure consisting of an upper and lower epidermis, between which are four inner layers which will produce the pallisade and mesophyll (Fig. 3 and 4). The numbers of cells in these six layers increases due to anticlinal division of the plate meristem over the whole area of the lamina. Dale (1976) suggests that the marginal and plate meristems might be less distinct than suggested by Maksymowych (1973). It is also clear that the view which was once held that leaf expansion was simply enlargement of cells previously produced by the primordia is incorrect. It has been shown for several species, e.g. Xanthium, tobacco and sunflower (Dale, 1982) that more than 90% of the cells of the leaf are produced after lamina expansion has begun. Towards the end of the cell division phase, periclinal divisions occur in the internal tissues of the leaf increasing the number of layers from four to seven or eight (Fig. 4). Cell division is complete in the lamina by LPI 3 and therefore all subsequent increase in leaf area must be due to cell expansion. Although Maksymowych (1973) makes a very rigid distinction between the phase of cell division and the later phase of cell expansion (Fig. 5), it is clear that cell volume stays constant during the phase of rapid cell division hence cell expansion must be occuring otherwise mean cell volume would decrease. The development of dicotyledon leaves takes on a further degree of complexity when the timing and duration of events is considered. It is known for several dicotyledon species that leaf growth ceases earlier at the tip of the leaf than at the base (Avery, 1935; Maksymowych, 1973; Dale & Milthorpe,

Table 1. Stages of leaf development in a dicotyledonous leaf

STAGE	LEAF PLASTOCHRON INDEX (LPI)	
Initiation of primordium	-7 to -6	Primordium 20-30 μm long at LPI-6.
Beginning of lamina formation	-4.8 to -5	Primordium 220 μm long 6 layers thick.
Marginal meristem active	-4.8 to +2.0	Leaf 0.25 - 55 mm long.
Plate meristem active	-4.0 to +2.8	Leaf 0.36 - 65 mm long.
Leaf unfolding begins	-0.5 to 0	Leaf 5 - 10 mm long.
	0	Leaf 10 mm long. Increase in thickness begins.
Plate meristem divides periclinally giving 7-8 cell layers	0.7 to 2.6	Lamina differentiation occurs first at the leaf tip.
Phase of rapid enlargement of lamina begins	2.0	Growth ceases first at the tip.
Cell division in lamina complete	2.5 to 3	Overestimate for cells contributing to lamina area.
Cell division in petiole complete	3.5	
Lamina expansion complete	6	Mature leaf 150-180 mm long, 120-150 cm^2 area
Petiole extension complete	7	

Data for *Xanthium pennsylvanicum*, derived from Maksymowych (1973). Leaf Plastochron Index (LPI) is a developmental scale. The zero point is taken when the leaf is 10 mm long. Leaves earlier in development than this have a negative LPI and later in development have a positive LPI. (Erickson & Michelini 1957). The time at which cell division ceases (LPI 2.5 to 3) might overestimate the contribution of cells which will contribute to lamina area, because at this time periclinal division of the plate meristem is occurring to increase the number of layers from the original four between the upper and lower epidermis to 7-8 layers.

1983), see also Fig. 4 and 6. It can be seen from Fig. 6 that adjacent regions of the leaf can have very different rates of relative growth. In addition the cessation of the cell division phase and the beginning of the maturation and expansion of the lamina begins first near the tip and continues much later in the basal portions of the leaf. The situation can arise (Fig. 4, LPI 2.30) where the tip of the lamina is almost completely mature, the middle portion of the leaf is undergoing periclinal cell divisions to produce the final number of cell layers, whilst at the base of the same leaf only the original six layered structure is evident. There is even heterogeneity at the tissue level. The cells of the palisade tissue increase mainly in height, whereas the adjacent epidermal cells increase in width. It has been generally assumed that it is the continued expansion in the epidermis which causes the inner cells of the leaf to separate and produce air spaces.

Consequences of the Complexity of Leaf Growth to Growth Substance Analysis in Leaves

Armed with this outline of leaf development and growth, it is apparent that studies of regulator involvement need to be sophisticated. Firstly it must be recognised that cell division and cell elongation are distinct phases, possibly with separate control mechanisms. The analysis of regulators possibly linked with cell division should be easier in grass leaves than in dicotyledonous leaves simply because the area of cell division is so well defined in the former. Grass leaves also present a good opportunity to study the relationship between growth substance concentration and rate of cell expansion because the zone of cell expansion is so distinct. It is apparent, however, that measuring the growth substance level in the whole leaf would be less likely to reveal any meaningful correlations. In the case of the maize leaf shown in Fig. 1, the leaf material most easily removed for analysis would not even be growing and regulators found there obviously could not be involved directly in controlling growth. The only area of the leaf of

Fig. 5. The relationship between DNA synthesis, cell division and cell enlargement (increase in cell volume) during leaf development in Xanthium. From Maksymowych (1973).

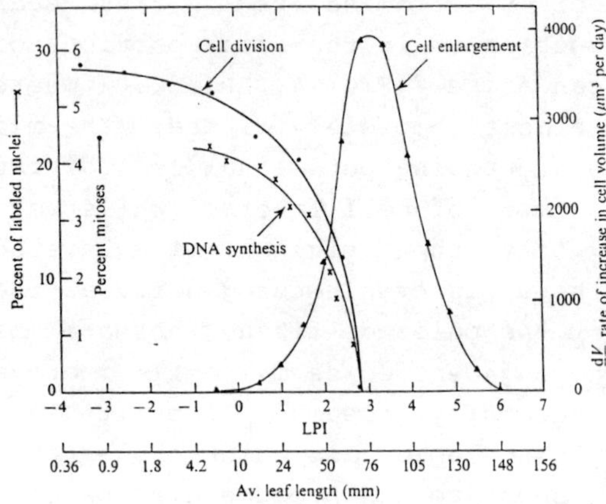

Fig. 6. Relative growth rates for expansion of different regions of Xanthium leaf nearing maturity (LPI 4.18). From Maksymowych (1973).

interest would have to be dissected out and would represent less than 10% of the lamina area. However, the fact that this clear pattern of growth exists in the developing grass leaf presents an exciting opportunity to those interested in the possible growth substance control of elongation. An analysis of these substances at positions along the leaf should reveal the correlations expected, if such correlations exist. In contrast to this attractive system in grass leaves, the heterogeniety of the dicotyledon leaf makes it a very difficult system for studying growth correlations. The gradients of growth rate are spatially complex hence the distribution of any regulators controlling expansion would have to be equally complex. Analysis at the whole leaf level is obviously inappropriate and at the very least the leaf would have to be analysed as to areas where there is a predominance of expansion and areas where little expansion is occuring. This fact was recognised in the early years of auxin research (Avery, 1935; Boysen-Jensen, 1936) and these workers not only analysed the auxin content of leaves of different ages but also assayed separately different areas of the leaf thought to be growing at different rates (Fig. 7). Interestingly, they reported marked differences in auxin concentration at different positions in the leaf and these studies should be repeated using more modern, definitive techniques. It is unfortunate that so many more recent studies of plant growth substances in leaves have treated the leaf as a homogeneous organ.

It is apparent that future attempts to correlate growth substance concentration with cell division or cell expansion in leaves should take the factors discussed above into account. However, another aspect of leaf development must also be considered - the temporal changes which occur throughout the day. If one measures leaf area of leaf length at infrequent intervals one might conclude that the growth rate of the leaf is fairly constant (Fig. 8). However, more frequent measurements of the rate of the leaf growth have shown that leaf growth varies very considerably during a daily cycle (Fig. 8 and 9). If plant growth

Fig. 7. Relative growth and polarised growth in a leaf of Nicotiana. Greatest growth is shown in the marginal and basal regions. Polarised growth (greater increase in length than width) is seen at the tip and the base. Lower diagrams (A, C & D) show growth-hormone, probably auxin, concentration in different parts of leaves of increasing age. Concentration expressed in plant units. After Avery (1935) in Boysen-Jensen (1936).

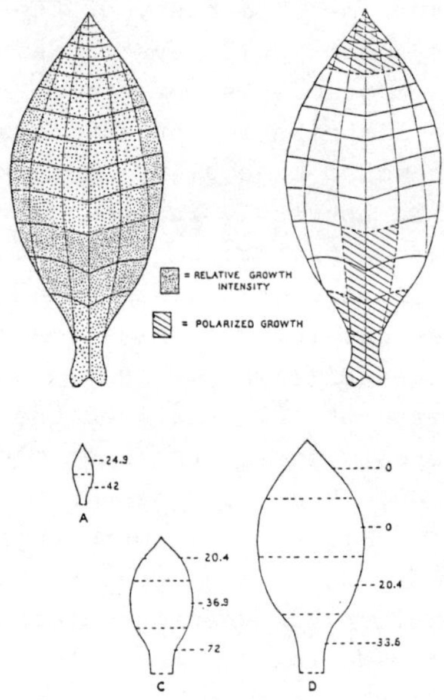

substances are controlling elongation it would obviously be necessary to measure them at different times during this cycle in order to establish the existence of meaningful correlations between leaf growth and growth substance levels.

CONCLUDING REMARKS

The idea that any one plant growth substance controls leaf development is not well supported by evidence and attempts to implicate more than one regulator have not

Fig. 8. The growth rate of sunflower leaves is not constant throughout the daily cycle. Infrequent measurements (right) fail to reveal the large diurnal growth rate changes (left). Note that diurnal changes are much less marked in the plants grown with high nitrogen. From Radin & Boyer (1982).

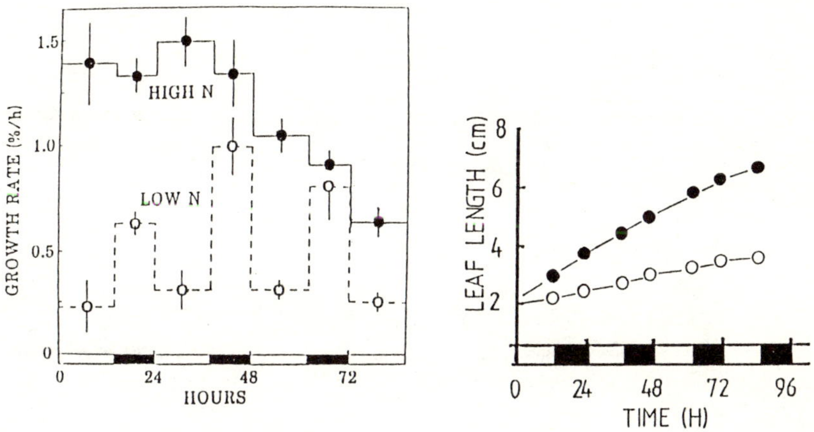

been much more convincing. It is the opinion of the authors that many attempts to correlate growth substance concentration changes and leaf growth have been of limited value because they have ignored the spatial and temporal heterogeneity of processes involved. Opportunities exist to improve on earlier work. For instance, the relatively simple developmental system in grass leaves remains to be exploited and deserves further study. The techniques for analysing small amounts of growth substances in limited amounts of tissue are now available and it should be possible in future to analyse individual growth substances in different parts of the leaf. However, it would be wise for those undertaking such studies in the future to dwell on whether growth substances need be evoked to explain any aspects of leaf development. As is made clear elsewhere in this volume,

Fig. 9. The daily pattern of growth of wheat leaves is shown by measuring the leaf length continuously. Note the rapid, short duration changes which occur after the light transitions in control leaves, (A). These transitions are largely abolished by treatment of the plant with ABA (B). From Christ (1976).

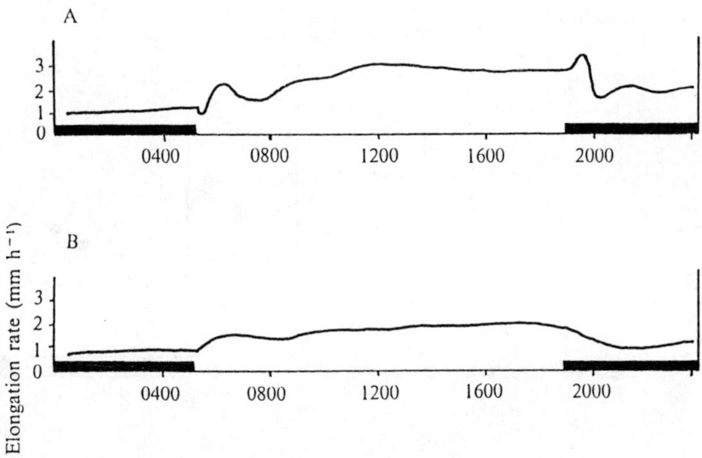

other factors such as water potential, nutritional status and light can have a profound effect on leaf development. There are reasons to believe that these factors can influence events directly and need not be growth substance mediated. Consequently, future attempts to implicate growth substances in leaf development should attempt to define a precise role for the substance under study. A hypothesis should be produced which outlines clearly why it is necessary to implicate the regulator, why other factors cannot account for the changes under study and how the proposed scheme of regulation fits into the known patterns of leaf development, both spatially and temporally. Although these might seem stringent demands, such precision is surely needed if we are ever to provide convincing answers to questions raised over the last half a century about the control of leaf growth by endogenous growth substances.

REFERENCES

Arber, A. (1950). The Natural Philosophy of Plant Form. Cambridge: Cambridge University Press.

Avery, G.S. (1935). Differential distribution of a phytohormone in the developing leaf of Nicotiana and its relation to polarised growth. Bull. Torr. Bot. Club, 62, 313-30.

Boffey, S.A., Selden, G. & Leech, R.M. (1980). Influence of cell age on chlorophyll formation in light-grown and etiolated wheat seedlings. Plant Physiol., 65, 680-84.

Boysen-Jensen, P. (1936). Growth Hormones in Plants. Translated and revised by G.S. Avery & P.R. Burkholder. New York: MacGraw-Hill.

Bradford, K. & Yang, S.F. (1980). Stress-induced ethylene production in the ethylene-requiring tomato mutant diageotropica. Plant Physiol., 65, 327-30.

Brenner, M.L., Hein, H.B., Schussler, J., Daie, J. & Brun, W.A. (1983). Coordinate control: the involvement of ABA, its transport and metabolism. In Plant Growth Substances 1983, ed. P.F. Wareing, pp. 343-52. London: Academic Press.

Christ, R.A. (1978). The elongation rate of wheat leaves. II. Effect of sudden light change on the elongation rate. J. Exp. Bot., 29, 611-8.

Cleland, R. (1983). Is plant development regulated by changes in the concentration of growth substances or by the changes in sensitivity to growth substances. Trends in Biochemical Science, 8, 354-57.

Dale, J.E. (1976). Cell division in leaves. In Cell Division in Higher Plants, ed. M.M. Yeoman, pp. 315-45. London: Academic Press.

Dale, J.E. (1982). The Growth of Leaves. Studies in Biology No. 137. London: Edward Arnold.

Dale, J.E. & Milthorpe, F.L. (1983). General features of the production and growth of leaves. In The Growth and Functioning of Leaves, ed. J.E. Dale & F.J. Milthorpe. pp. 151-78. Cambridge: Cambridge University Press.

Erickson, R.O. & Michelini, F.J. (1957). The Plastochron index. Amer. J. Bot., 47, 350-51.

Esau, K. (1953). Plant Anatomy. New York: Wiley.

Goodwin, P.B. (1978). Phytohormones and the growth and development of organs of the vegetative plant. In Phytohormones and Related Compounds: A Comprehensive Treatise, Volume II, ed. D.S. Letham, P.B. Goodwin & T.J.V. Higgins, pp. 94-109. Amsterdam: Elsevier/North Holland.

Goodwin, P.B. & Erwee, M.G. (1983). Hormonal influences on leaf growth. In The Growth and Functioning of Leaves, ed. J.E. Dale & F.L. Milthorpe, pp. 207-32.

Cambridge: Cambridge University Press.
Gregory, F.C. (1928). Studies in the energy relation of plants. II The Effect of temperature on increase in area of the leaf surface and in dry weight of Cucumis sativus. Ann. Bot. 42, 469-507.
Hsiao, T.C. (1973). Plant responses to water stress. Ann. Rev. Plant Physiol., 24, 519-70.
Jackson, M.B. (1979). Is the diageotropic tomato ethylene deficient? Physiol. Plant, 46, 349-51.
Maksymowych, R. (1973). Analysis of Leaf Development. Cambridge: Cambridge University Press.
Michelena, V.A. & Boyer, J.S. (1982). Complete turgor maintainance at low water potentials in the elongating regions of the maize leaves. Plant Physiol., 69, 1145-49.
Phinney, B.O. & Spray, C. (1983). Chemical genetics and the gibberellin pathway in Zea mays L. In Plant Growth Substances 1983, ed. P.F. Wareing, pp. 101-10. London: Academic Press.
Potts, W.C. & Reid, J.B. (1983). Internode length in Pisum III. The effect and interaction of the Na/na and Le/le gene differences on endogenous gibberellin like-substances. Physiol. Plant Physiol., 57, 448-54.
Radin, J.W. & Boyer, J.S. (1982). Control of leaf expansion by nitrogen nutrition in sunflower plants. Plant Physiol., 69, 771-75.
Sharman, B.C. (1942). Developmental anatomy of the shoot Zea mays L. Ann. Bot. 6, 245-82.
Thimann, K.V. (1977). Hormone Action in the Whole Life of Plants. Amherst: Massachusetts Press.
Trewavas, A.J. (1983). Is plant development regulated by changes in the concentration of growth substances or by the changes in sensitivity to growth substances. Trends in Biochemical Sciences, 8, 354-57.
Van Volkenburgh, E. & Davies, W.J. (1983). Inhibition of light stimulated leaf expansion by ABA. J. Exp. Bot., 34, 835-45.
Went, F.W. (1951). The development of stems and leaves. In Plant Growth Substances, ed. F. Skoog, pp. 287-98. Madison: University of Wisconsin Press.
Wright, S.T.C. (1978). Phytohormones and stress. In Phytohormones and Related Compounds: A Comprehensive Treatise, Vol. II, ed. D.S. Letham, P.B. Goodwin & T.J. Higgins, pp. 495-532. Amsterdam: Elsevier/North Holland.
Zeevaart, J.A.D. (1977). Sites of ABA synthesis and metabolism in Riccinus communis L. Plant Physiol., 59, 788-91.
Zobel, R.W. (1973). Some physiological characteristics of the ethylene requiring tomato mutant diageotropica. Plant Physiol., 52, 385-89.

A PIVOTAL ROLE FOR NITRATE AND LEAF GROWTH IN PLANT DEVELOPMENT

A. Trewavas
Department of Botany, University of Edinburgh,
Edinburgh, EH9 3JH, United Kingdom.

INTRODUCTION

That the processes of development involve communication seems self evident. Yet the means whereby that communication occurs is the subject of considerable debate. It is the function of this article to point to growing leaves and nitrate as playing a possible pivotal role in whole plant communication particularly as regards photoperiodism. The article discusses initially aspects of communication, then considers leaf growth and nitrate and then places this in the context of a number of photoperiodic phenomena.

THE PROBLEM OF COMMUNICATION IN PLANTS

It is easy to show that the various parts of a growing plant interact with each other. These interactions, misleadingly called "correlations", are obvious evidence of communication. There are two major ways in which this communication can be considered to occur.

The first of these supposes communication to result from the movement of specific chemicals from a site of synthesis to a site of effect; that is, communication occurs in a positive way by means of hormones. Changes in hormone concentration it is believed can then manipulate the size of the response. Based, by and large, on analogies with mammalian hormones, this hypothesis has invested the five major growth substances (and other ill-defined substances such as florigen, tuberin, dormin *etc*.) as being the hormones conveying information from one part of the plant to another. Although initially credible when first proposed in the 1930's, this hypothesis is now known to have serious defects (Trewavas 1981, 1982a, b, 1983a).

There are many sites of synthesis of growth substances and virtually every stage of plant development has been shown to be sensitive to at least four if not the five separate growth substances. Meaningful communication requires a localisation of both synthesis and response since the affected tissue cannot otherwise detect the tissue of origin nor can the site of synthesis determine what has been affected. To complicate the situation, many if not all processes of development have been shown to be modified, accelerated or induced by a wide variety of organic and inorganic chemicals and environmental factors as well as growth substances. The responsiveness of tissues to growth substances is thus greatly variable. Frequently the sites of synthesis of growth substance and effect appear to be identical thus apparently obviating a messenger function. For example, the cereal aleurone and elongating pea stem cells synthesise their own biologically-active gibberellins (Atzhorn and Weiler, 1983; Ingram et al., 1984). In any one plant there is great variation throughout the year in the numbers of supposed rich tissue sources of growth substances (young leaves and seeds) and growth substances are released into and circulate throughout a plant body which is both plastic and indeterminate in volume and in a vascular system whose composition and flow rate vary greatly from hour to hour. None of this suggests that plants can accurately regulate circulating growth substance levels and dose response curves to growth substances (generally over four orders of magnitude) suggest the system is designed to tolerate variation rather than be sensitive to it.

Growth substances are important in plant development not least because they permit the expression of groups of genes concerned with specific aspects of growth, but also because they function in the co-ordinated growth responses of different cell types in growing areas. They are probably also responsible for smoothing out growth which must occur with variable input of substrates, i.e. the metabolic fly wheel concept (Trewavas, 1983b). Communication is not essential to these functions.

The second approach to communication in plants takes cognisance of the specific characteristics of plant construction and growth and argues instead that competition between growing areas is the major form of communication.

That the various growing or metabolically active parts of a plant compete with each other for available circulating growth materials has been well-established and was well known even in the last century. Goebel (1900) quotes a number of examples and many papers since that time have established the point very clearly. The population ecologists have recently rediscovered the important concept of the plant as a population of growing, competing meristems analogous to colonial coelenterates (White, 1978).

Such a concept can be generalised to all metabolically-active plant parts of which growing leaves represent a major constituent. Virtually all higher plants are constructed from repetitions of a basic organising module structure. In the shoot the module consists of a bud plus subtended leaf and in the root repetitions of root meristems. Growth takes place by the addition of new modules to the pre-existing structure. Plants possess no obvious centralised controlling tissue and the individual meristems (and thus growing leaves) can be demonstrated to a large extent to grow and develop autonomously. Such autonomy must form the basis of the well-characterised "opportunistic growth" of plants (Tomlinson, 1982) and in part account for the remarkable semi-autonomous character of individual plant parts exemplified by physiological regeneration. A system of competition and partial independence is almost the only way in which sensible organisation of many angiosperms could occur. However the capability for growth of any growing tissue will depend on the availability of specific growth materials, the metabolic activity of the growth area and the capability for transporting such materials into the growing region.

There are two features of this competitive model

which are not widely appreciated. Firstly, what one growing area removes from circulation is not available to another. This represents a form of communication between the two areas which may be termed negative control and would be very difficult to distinguish, except with further analysis, from a positive hormonal model. Secondly,since the different growing areas (e.g. leaves compared to meristem) will have qualitatively different growth material requirements, reducing the growth rate (competitive ability) of some will result in qualitative changes in what is left to circulate in the vascular system. Such qualitative changes could then at as effective signals for developmental change.

Negative controls as a means of changing development do not seem to be widely appreciated although they can account for flowering, apical dominance, fruit set, abscission and root-shoot ratios (Sachs, 1977). Circulating growth substances too will be subject to competition and an increase in binding or transporting proteins in one growing tissue would sequester more from the general circulation. There is no doubt that negative control more accurately fits the novel growth characteristics of plants and explains much more simply and easily interactions between plant tissues.

NITRATE AND LEAF GROWTH

Young, expanding, leaves are one of the major growing tissues in any plant and also act as critical sources of growth materials such as carbohydrate and organic nitrogen. They must therefore figure prominently in any consideration of negative controls and development.

Leaves require a great variety of factors for growth but they have long been known to be especially sensitive to nitrate, the most abundant form of nitrogen in the soil. Nitrate has more effect on leaf growth than any other major nutrient. Increased nitrate levels can dramatically accelerate leaf elongation rates and length and the final area of individual leaves may be increased over three fold (Dale, 1982; Terry *et al*., 1983). The dose response curve for nitrate has been carefully measured and ranges over little more than an order of magnitude in concentration; leaf growth rates may be increased by nearly ten fold as plants experience the range of limiting to optimal nitrate treatments (Ingestad and Lund, 1979). This should be compared with the generally much wider dose response curves found for growth substance effects. Apical dominance may be reduced or completely broken by nitrate treatment and thus many more leaves initiated into the growth process. The effects of applied nitrate are thus basic, dramatic and fundamental in leaf growth and expansion.

Despite the dramatic effects of nitrate on leaf growth, the mechanisms underlying this phenomenon seems to have attracted remarkably little attention. It has been shown that nitrate application may lead to increases in both cell size and number but it is not known whether this is a direct or indirect effect on mitosis or cell expansion. Chloroplast numbers per cell may be increased and there may be higher levels of chlorophyll and changes in plastid fine structure. With a lack of nitrate, leaf senescence may be accelerated as nitrogen is withdrawn and exported from the older to younger leaves.

Why is leaf growth so sensitive to nitrate? One reason is probably that leaves have a much higher nitrogen content than other vegetative plant tissues on a dry or fresh weight basis. Thus the leaves of the potato have a three fold higher nitrogen content than young growing tubers or stems or roots (Harris, 1978); both flowers and dormant buds have much lower contents of extractable protein than leaves. The difference between leaves and these other tissues seems to be a requirement for substantial amounts of protein for construction of the photosynthetic apparatus. This does mean that in times of nitrogen shortage the growth of leaves may be impaired to a much greater extent than that of tubers, flowers or other vegetative tissues. The exception to this rule may be developing seeds and Stoddart and Thomas (1983) have pointed to a possible interesting relationship between seed nitrogen content and associated leaf senescence.

It is well known that leaves and roots contain nitrate reductase and although some reduced organic nitrogen may be acquired from the roots, nitrate has been detected in xylem sap. It could be anticipated that growing leaves may synthesise their own reduced nitrogen or acquire it from leaves close by. In the early stages of growth, evidence suggests that the young leaf depends primarily on export of nitrogen from older leaves. As it continues to grow an increasingly greater proportion is derived from the roots and by the time the leaf is mature there is a constant import and export of nitrate and other nitrogenous materials. Because growing leaves are heavy consumers of nitrate and reduced or organic nitrogen, environmental conditions which substantially alter leaf growth should lead to specific modifications in the inorganic and organic materials circulating in the vascular system.

PHOTOPERIODIC CONTROL AND A CENTRAL ROLE FOR NITRATE AND LEAF GROWTH IN DEVELOPMENT

The C/N Ratio

Sucrose and nitrate or reduced nitrogen are two of the basic materials found in the vascular system and which are essential for growth and development. They are two of the materials available for competition between the metabolically-active areas of the plant. Much work in the earlier part of this century suggested that the ratio of carbohydrate to nitrogen (the C/N ratio) be critical to many aspects of plant development. The seedling using its nitrogenous reserves has a low C/N ratio. This promotes shoot growth because of the high nitrogen requirement for growing leaves. As the photosynthetic area increases and with a more rapid growth of the shoot the C/N ratio continues to increase until a stage of maturity is reached at which specialised acts of development can now commence and different tissues, *e.g.* flowers and tubers, whose composition is likewise high in C/N, appear (Loomis, 1953; Sinnott, 1960). With a high circulating C/N ratio, root growth is promoted because the root competes more effectively for the nitrate it is absorbing and because compositionally it is high in C/N. With a low C/N ratio the shoot competes more effectively for its own photosynthate and thus can grow more quickly. The idea is a simple one for which there is substantial supporting evidence but has never received a thorough going examination with modern technology. It is time that it did.

Table 1 summarises a wealth of more recent data which suggests a relationship between photoperiodic control, the C/N ratio and specific acts of developmental change such as flowering, tuberisation, dormant bud formation, etc. The table shows that in many plants development can be changed and photoperiodic control subverted either by artifically elevating sucrose levels or by lowering available nitrogen or nitrate (i.e. increasing C/N ratio). As a corollary to the latter observation, application of nitrate to plants generally inhibits flower and tuber induction, rhizome formation, etc. and associated with this inhibition is a corresponding promotion of leaf growth. The application of nitrate will of course lower the C/N ratio. Induction of flowers or tubers is usually associated with a pronounced reduction in vegetative and leaf growth. Alternatively, for example, in the case of dormant tree buds, application of nitrate late in the year can lead to their outgrowth. Further discussion of the data in this table can be found in Trewavas (1983b).

Changes in the C/N ratio as a means of explaining the induction of flowering, tuberisation or dormancy are found in the older literature but fell into disrepute. It is pertinent to inquire into the reasons for this and there seem to be three major ones. First, most investigators assumed that an increase in carbohydrate was sufficient for induction and the notion that N should also be lowered was not considered nor at the time were mechanisms available to explain how this might happen. Second, the theory was tested in a far too simplistic way with assumptions made concerning elevation or lowering of C or N from experimental

Table 1. Developmental phenomena altered by balance of sucrose to nitrate or reduced nitrogen. Adapted from Trewavas (1983b).

Developmental response	Photoperiodic control	Vegetative growth reduced during induction	Response induced by sucrose application	Response inhibited by increased nitrate application and induced by lowered nitrate availability	Growth substance supposedly involved
Flowering induction	Yes	Yes	Yes	Yes	Florigen? Gibberellin?
Tuber induction	Yes	Yes	Yes	Yes	Kinin? Tuberin?
Rhizome formation	Yes	Yes	Yes	Yes	Not known
Dormant bud (turion) formation	Yes	Yes	Yes	Yes	Dormin? Abscisic acid?
Leaf senescence and abscission	Yes	Yes	?	Yes	Cytokinin Ethylene/auxin
Reduction in leaf growth	Yes	Yes	Yes	Yes	-
Xeromorphic characters	Not known	Yes	Yes	Yes	Abscisic acid?

manipulations without the necessary accompanying analysis. The possibility of enzymatic variation between different plant tissues and organs in the production of carbohydrate and organic nitrogen was not considered or understood. Third, there were some notable exceptions to a simple application of the hypothesis. For example, nitrate was found to promote flowering in *Xanthium*. This will be referred to later because *Xanthium* is also somewhat exceptional in its nitrogen metabolism.

As a consequence it became supposed that photoperiodic control was mediated by specific hormones such as florigen, tuberin, dormin and the C/N hypothesis was largely abandoned. But as Sinnott (1960) points out the theory suggests development to be controlled as a balance of promotors (C) to inhibitors (N), a concept still strongly held today.

Nitrate Reduction and Leaf Growth are Pivotal to the Control of the C/N Ratio

There seem to be two linking characteristics in the phenomena in Table 1. Firstly, all the systems involve growing tissues with a demonstratable interaction (competition) between the various growing parts, such as the leaves and shoot apex which are involved in the induction process. The leaves most sensitive to inductive photoperiods for flowering seem to be the rapidly-expanding ones for example and it is the shoot apex which produces flowers. Secondly, the induction processes seem to involve and conversely, can be inhibited by changing the balance of growth amongst the constituents; most critical of these again are the growing leaves. Induction is frequently associated with the inhibition of leaf growth; promotion of leaf growth by nitrate inhibits photoperiodic induction.

A simple hypothesis therefore to explain the photoperiodic induction of flowers, tubers, rhizomes or dormant buds suggests that the initial metabolic change instituted by inductive light periods is to decrease the synthesis of organic nitrogen. Since leaves are the light period sensors it is here that the metabolic change must be instituted. Leaf growth consequently should be impaired because it is so sensitive to organic nitrogen and the carbohydrate normally used by the growing leaves would be released into the circulation. The C/N ratio would be elevated and sensitive cells in the shoot, stolon apex or elsewhere would respond accordingly. The effect of the inductive photoperiod would be to reduce the ability of the leaf to compete for the available carbohydrate and organic nitrogen resources.

Nitrate reductase is usually considered to be the limiting enzymatic step in organic nitrogen synthesis and the activity of this enzyme has been reported to be under photoperiodic control (Srivastava, 1980). Is this the critical photoperiodically controlled enzyme? It is demonstratably not as simple as that. Individual plant species vary greatly in the proportion of nitrate reductase to be found in leaves or roots. Some plants, most notably *Xanthium*, have no nitrate reductase in the roots at all or organic nitrogen in the xylem sap (Carr, 1966). In *Xanthium*, inductive photoperiods can initially promote leaf growth although this is normally followed by later pronounced inhibition (Bernier *et al*., 1981). This complex of behaviours can however still be shown to be consistent with an elevated C/N ratio by taking account of what happens in the rest of the plant.

Consider two different plants A and B. Plant A has abundant nitrate reductase in the roots and leaves as do many plants; plant B on the other hand has nitrate reductase only in the leaves, like *Xanthium*. If plant A is exposed to inductive photperiods and leaf nitrate reductase lowered, the root system would continue to synthesise its own organic nitrogen and thus sequester carbohydrate from the leaves. Root growth would thus continue, the C/N ratio would rise and leaf growth would be impaired.

In plant B however the root system is dependent on the leaves for both carbon and organic nitrogen. With a lowering of leaf nitrate reductase, consequent on induction, root growth would be severely impaired or even cease, since leaves would have the first call on the more limited amount of organic nitrogen they still synthesised. It is possible to envisage under these conditions that leaf growth might continue at the same rate or even be promoted, particularly if leaf growth had previously been somewhat limited by the necessity for providing carbohydrate for the growth of the root system. However it could be anticipated that this initial promotion would be circumscribed by the lack of concurrent root development and inhibition of leaf growth would eventually result. Other growing areas, particularly the shoot apex, would still experience an elevated C/N ratio because of the lowering of nitrate reductase and a reduction in circulating organic nitrogen. Under extreme situations nitrate application might well promote flowering because of insufficient organic nitrogen for the floral apex to grow.

It should be obvious from these two extreme examples that a range of leaf growth behaviours could be expected, dependent on the amounts of nitrate reductase in roots or leaves and the degree of inhibition of nitrate reductase. Additional variables complicating the situation could include genotype or cultivar enzyme variation, previous light and nitrogen treatment, possible storage of nitrate or organic nitrogen in vacuoles, nitrate treatment at time of induction, light intensity during induction and so on. However a range of leaf growth behaviours is what is observed. In many plants leaf growth inhibition accompanies photoperiodic induction but in *Xanthium* and a few others leaf growth continues or may even be promoted. This is usually followed by pronounced inhibition and sometimes changes in leaf shape and character suggesting that the fundamental balance of necessary growth materials has altered. Such considerations indicate the complex picture which follows from simple models of competition amongst the different growing areas of the plant.

THE CENTRAL ROLE OF LEAF GROWTH IN DEVELOPMENT

Leaf growth occupies a central role in development.

Not only is the expansion and unfolding of leaves crucial to early seedling establishment but changes in growth later on seem to be responsible for the initiation of specific acts of development. The argument applied here is that with the known sensitivity of leaf cell division and expansion to nitrate, photoperiodic control of organic nitrogen synthesis is the simplest way of manipulating subsequent growth. The net result of inductive photoperiods is an increase in available carbohydrate, which would normally have been used instead by the growing leaf and a lowering of available organic nitrogen.

Perhaps the most critical element in this article is that it shows that there are two ways of interpreting apparent communication in plants. The first, the hormonal model, exemplified by florigen argues for a direct, specific chemical signal moving from leaf to shoot apex. It focusses attention on only two constituents of an obviously complex system. The second, competitive, model points to systems of communication as consisting of competitive interactions between tissues which have different growth requirements (Sachs, 1977). Upsetting or changing the balance of available growth materials impairs the growth of one tissue, the growing leaf, thereby enabling others, e.g. the shoot apex, to sequester a greater share of the now altered resources. These observations suggest that changes in the balance of growth amongst the various plant tissues to be crucial aspects of altered development. The change in the C/N ratio may both induce and permit the growth of tissues which compositionally have a high C/N ratio.

This latter model is preferred here because it is more compatible with the modular, competitive model of negative control in plants outlined in the introduction. It thus fits more closely the known and novel aspects of plant growth. It explains why, despite numerous attempts, substances like florigen have never been found. This is because the induced leaf is an essential component of the system; an interaction between leaves and shoot apex is essential for flowering to occur.

There are similarities between the hypothesis proposed here and the nutrient diversion theory of flowering proposed by Sachs (1977). Sachs argued that the shoot apex and young leaves compete for available circulating carbohydrate. Changing the growth rate of the young leaves led to nutrient diversion to the shoot apex which then flowered. In this hypothesis, the youngest leaves, since they are nearest the shoot apex, exert a disproportionate effect on apical development. Critically, Sachs (1977) points out that nutrient diversion is indistinguishable at a gross level from a florigen hypothesis; i.e. negative interactions are superfically indistinguishable from positive controls. Sachs (1977) also lists a variety of other phenomena such as fruit set, abscission, root:shoot ratio etc. where changes in balance in an interacting system of leaves and other tissue seem critical.

Taking account of more components of the system may help explain other apparent communication phenomena which are in equal difficulties with the identity of the supposed signal, such as apical dominance. Investigations of apical dominance have always concentrated on the shoot apex and the dormant lateral bud; other possible interactions with other tissues seem to have received little attention. Yet careful experimentation has shown that apical dominance can be broken by curtailing the growth of or simply removing young growing leaves (Hillman, 1984). Alternatively, providing an excess of nitrogen can break dominance in some plants, a treatment which it could be anticipated would selectively promote leaf growth. These data is suggestive again of a form of competition between growing leaves and buds which is mediated by nitrate. An excess of this compound reduces the competition and permits maximal leaf and bud growth. Removal of the growing leaves reduces competition for a limited resource. Again the phenomenon involves a change in balance of growth amongst the growing, interacting constituents.

We can finish here by again questioning the underlying assumption of the necessity of specific substances such as florigen, tuberin or dormin for specific developmental control. It is argued that such substances are necessary to induce the expression of specific genes in the sensitive cells which then result in altered development. But what are these genes? Examining the early changes in a shoot or stolon apex or a dormant bud after induction one cannot help be struck by the fact that the initial events seem to be merely changes in the direction of cell division and growth, i.e. changes in the balance of genes which are already being expressed. Only later do recognisably novel genes appear to be translated. Such observations are entirely consistent with the argument developed here concerning changes in the balance of C/N in induction and the central role of leaf growth and nitrate in these processes.

REFERENCES

Atzhorn, R. & Weiler, E.W. (1983). The role of endogenous gibberellins in the formation of amylase by aleurone layers of germinating barley caryapses. Planta 159, 289-99.

Bernier, G., Kinet, J.M. & Sachs, R.A. (1981). The Physiology of Flowering. Florida: CRC Press.

Carr, D.J. (1966). Metabolic and hormonal regulation of plant growth and development. In Trends in Plant Morphogenesis, ed. E.W. Cutter, pp. 253-87. London: Longmans.

Dale, J. (1982). The Growth of Leaves. London: Edward Arnold.

Goebel, K. (1900). Organography of plants. Volume 1. Reprint 1969. New York: Hafner Publishing Co.

Harris, P.M. (1978). Mineral nutrition. In The Potato Crop, ed. P.M. Harris, pp. 195-244. London: Chapman and Hall.

Hillman, J. (1984). Apical dominance. In Advanced Plant Physiology, ed. M.B. Wilkins, pp. 127-48. London: Pitman Publishing Ltd.

Ingram, T.J., Reid, J.B.,Murfet, I.C., Gaskin, P., Willis, C.L. & MacMillan, J. (1984). Internode length in Pisum. The le gene controls the 3B hydroxylation of gibberellin A to gibberellin A . Planta 160, 455-64.

Ingestad, T. & Lund, A.B. (1979). Nitrogen stress in birch seedlings. Physiol. Plant. 45, 137-48.

Loomis, W.E. (1953). Growth and differentiation in plants. Iowa: Iowa State College Press.

Sachs, R.M. (1977). Nutrient diversion; an hypothesis to

explain the chemical control of flowering. Hort. Science 12, 220-22.
Sinnott, E.W. (1960). Plant Morphogenesis. New York: McGraw Hill.
Srivastava, H.S. (1980). Regulation of nitrate reductase activity in higher plants. Phytochem. 19, 725-33.
Stoddart, J.L. & Thomas, H. (1982). Leaf Senescence. In Encyclopedia of Plant Physiology, New Series, Volume 14A, pp. 592-636. Berlin: Springer-Verlag.
Terry, N., Waldron, L.J. & Taylor, S.E. (1983). Environmental influences on leaf expansion. In The growth and functioning of leaves, ed. J.E. Dale & F.L. Milthorpe, pp. 179-205. Cambridge: Cambridge University Press.
Tomlinson, P.B. (1982). A morphogenetic basis for plant morphology. In Axioms and Principles of Plant Construction, ed. R. Sattler, pp. 162-84. The Hague: Martinus Nijhoff.
Trewavas, A.J. (1981). How do plant growth substances work? Plant, Cell and Environment 4, 203-28.
Trewavas, A.J. (1982a). Growth substance sensitivity: the limiting factor in plant development. Physiol. Plant. 55, 60-72.
Trewavas, A.J. (1982b). The regulation of development and its relation to growth substances. Whats New in Plant Physiology 13, 41-4.
Trewavas, A.J. (1983a). Plant growth substances; metabolic fly wheels for plant development. Cell Biology International Reports 7, 569-575.
Trewavas, A.J. (1983b). Nitrate as a plant hormone. In British Plant Growth Regulator Group Monograph 9, ed. M.B. Jackson, pp. 97-110. British Plant Growth Regulator Group.
White, J. (1978). The plant as a metapopulation. Ann. Rev. Ecol. and System. 10, 109-145.

THE SYNTHESIS OF CELLULAR COMPONENTS IN LEAVES

Rachel M. Leech
Department of Biology, University of York,
Heslington, York, YO1 5DD, United Kingdom.

INTRODUCTION

Leaf development is remarkably consistent so the majority of higher plants can be instantly recognised by the shape of their leaves. Leaf tissue is heterogeneous and composed of a myriad of individual cells but the tight correlation between the patterns of development of the millions of constituent cells and their components ensures the consistency of leaf development. The internal cellular controls are numerous and complex, the constraints constantly changing during leaf development and regulation can occur at every stage during the synthesis of each cellular component. Controls are likely to operate wherever there is a biochemical limitation and this may be during transcription, translation, processing and/or assembly of the cellular components. Although the functions of the components of the mature leaf cell are still incompletely understood and we have only a fragmentary understanding of the controls which operate during their development, the challenges presented by the study of the molecular and cellular aspects of leaf development have already stimulated several detailed research programmes. The results of these studies relate almost exclusively to aspects of the synthesis of chloroplast components but provide graphic illustrations of the nature and complexity of the interactions which operate within a developing leaf cell.

LEAF MATERIAL - CHOICE AND MODE OF ANALYSIS

The study of the metabolic aspects of leaf development is experimentally difficult since procedures are needed which allow the simultaneous examination of populations of cells which differ both morphologically and functionally and develop at different rates within the same leaf over periods of days or weeks: epidermal tissue, mesophyll cells and veinal tissue all have different developmental patterns. The plant species must be chosen with care, the environmental changes during development reduced to a minimum and the basis for the comparison of the developmental changes carefully considered. It has been found that the units of expression which provide the best insight into developmental changes are those which are *biologically relevant* and so if the individual leaf or cell or molecule is chosen as the reference parameter, the results from the analysis of episodically sampled tissue can be most unequivocally interpreted. Cellular metabolic changes expressed in terms of gross leaf parameters such as dry weight, fresh weight, nitrogen or chlorophyll content are usually impossible to interpret since every feature of a developing leaf is continually changing.

The cell biologist studying leaf development has an apparent choice, *i.e.* to examine either a dicotyledonous or a monocotyledonous leaf. However, dicotyledons are unsuitable, except for limited morphological observations, and reference should be made to the excellent monograph of Maksymowych (1973) which graphically describes the morphological complexity of leaf tissue development in the dicotyledon *Xanthium* (cocklebur). This will convince any cell biologist of the very difficult sampling problems which need to be resolved before such a heterogeneous leaf tissue can be usefully employed in biochemical studies. There are also considerable logistic difficulties involved in obtaining sufficient leaf material from the dicotyledonous apical meristem for the biochemical analysis of the earliest stages of cell development.

In contrast the leaves of monocotyledonous grass species are particularly suitable for studies of leaf development. In these leaves cell division is restricted to a basal meristem so that each leaf provides a linear gradient of cell development from its base to its tip. In grass leaves leaf slices sectioned at right angles to the long axis contain cells and organelles of increasing maturity so the dynamics of their development can be investigated and also related to cell age. For these reasons monocotyledons are now increasingly used in developmental studies. Leaves of grasses have been used successfully for several developmental studies of leaf components and the species include wheat (Boffey *et al.*, 1979, 1980; Boffey & Leech, 1982; Baker *et al.*, 1983, 1984) ryegrass (Brangeon & Mustardy, 1979) maize (Hawke *et al.*, 1974; Baker & Leech, 1977; Chapman & Leech, 1979; Miranda *et al.*, 1981a, b) barley (Viro & Kloppstech, 1980; Leech & Walton, 1983; Nivison & Stocking, 1983) and oats (Taylor & MacKender, 1977). Most of the work discussed in this article relates to grass species showing this type of leaf development.

CELLULAR COMPONENTS - METHODS OF ANALYSIS

The earliest biochemical studies of cellular changes in developing leaves were descriptive and recorded large changes in classes of molecules such as proteins, nucleic acids, lipids and pigments (see Leech & Baker, 1983, for a review). All these components are synthesised and accumulate rapidly during leaf expansion and quantitative changes have been measured in a variety of species and under differing environmental conditions. It is not possible however to relate many of these measurements to specific steps in cellular development because they were made using dicotyledonous leaves in which a mosaic of groups of cells with differing patterns of cell expansion are adjacent to each other and it is now recognised that a developing dicotyledonous leaf cannot be sampled in such a way that either cell division or cell expansion is totally excluded (Scott & Possingham, 1983). Changes in cellular components measured by sampling progressively older leaves reflect the totality of all the changes in the mosaics of cells developing at different rates. The organelles in adjacent cells may differ profoundly, *e.g.* in levels of protein synthesis or in photosynthetic competence. Only 50% of the leaf cells in most leaves are mesophyll cells: the rest are epidermal cells and cells of the veinal tissue (Jellings & Leech, 1982).

In dicotyledons, in which meristematic activity occurs throughout the plate meristem throughout expansion (Dale, 1981), cytological and cytochemical investigations of the component changes in developing cells *in situ* within the leaf are the least equivocal method of determining the characteristics of individual cells. In a classical cytological study Maksymowych and Kettrick (1970) determined the timing of cell division, nuclear DNA synthesis (using ^{3}H-thymidine incorporation) and cell enlargement patterns during the early stages of leaf expansion in *Xanthium*. Immunocytochemical techniques offer a powerful tool to examine intracellular changes in such heterogeneous leaf tissue, but have yet to be exploited in studies of developing leaves as they have been in the demonstration that ribulose bisphosphate carboxylase (RuBPCase) is restricted to bundle sheath chloroplasts in mature leaves of C4 plants (Hattersley *et al*., 1977).

THE DEVELOPMENT OF CHLOROPLASTS IN MESOPHYLL CELLS

The changes in the components of the mesophyll cell during leaf development have been the most extensively investigated. The general strategy adopted has been an analytical one, that is to quantify the changes in individual cellular constituents in the anticipation that by assembling many observations it will be possible to describe more comprehensively the changes which occur during cellular differentiation. The development of organelles occurs concurrently with mesophyll cell enlargement but the chloroplast is the only mesophyll cellular organelle which has been examined in detail during leaf development. This is perhaps not surprising since the plastid is the organelle which changes most dramatically during leaf cellular development.

The major phase of chloroplast development occurs

after cell division, during and after the phase of rapid cell expansion. During leaf development, increases in mesophyll cell size and in chloroplast size often parallel each other and, because the chloroplasts also replicate, the proportion of the cytoplasmic volume occupied by chloroplasts increases more rapidly than either cell or chloroplast volume. As a result, in non-limiting environmental conditions, the cell and cytoplasm become completely packed with chloroplasts. The chloroplasts lie in a monolayer just inside the cell wall and although smaller cells have fewer chloroplasts the mesophyll cells within a leaf show a remarkable uniformity in the packing of their chloroplasts. The number of chloroplasts per cell is genetically determined and in wheat each genotype can be identified by its mean chloroplast number. The largest differences are found in species of different ploidy (Butterfass, 1979, 1983) and in wheat each doubling of the chromosome number is accompanied by an approximately 80% increase in mean chloroplast number (Dean & Leech, 1982b). The relationships between cell and chloroplast volume found in cells of increasing age within a young wheat leaf are illustrated in Fig. 1.

Chloroplast Growth

As a result of recent detailed ultrastructural and biochemical investigations we now have a rather comprehensive picture of both the structural changes and the quantitative changes in chloroplast DNA (cpDNA) and nuclear DNA (nDNA) which occur during plastid and cell differentiation. The proplastid to chloroplast transition involves changes in plastid size, shape and number and the sequence of events seems to be generally similar in plastids of young *Phaseolus* (French bean) leaves sampled at increasing age (Whatley, 1979) and in plastids of increasingly older cells within a single leaf of a grass such as rye (Brangeon & Mustardy, 1979) or maize (Whatley, 1977). In leaf meristematic cells the plastids (proplastids) are small (less than 1μm in diameter), pleiomorphic but

Fig. 1. Changes in cell, plastid area and total plastid area during cellular development within the first leaf of Triticum aestivum cv. Maris Dove. Values for cell and plastid areas are means for 100 measurements. Changes in cell and plastid shape and size are also indicated.

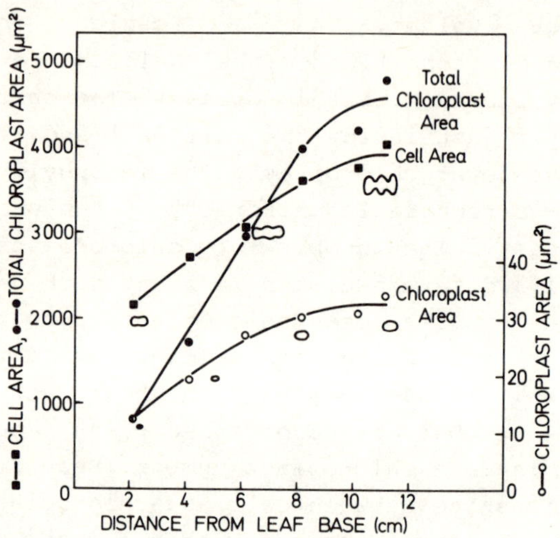

often spherical and bounded by a double membrane. The internal membranes are sparse, generally concave and perforated and contain only traces of chlorophyll. As cellular development proceeds, the differentiating plastids pass through a sequence of stages of starch accumulation and loss, an amoeboid stage and a major period of internal membrane proliferation. Thylakoid membrane biogenesis is massive, rapid and well ordered. In-folding in the regions where the membranes are perforated leads to the ordered spiral stacking of the membranes into grana which are characteristic of many angiosperm chloroplasts. In most species which have C4 photosynthetic mechanisms, one type of plastid-containing tissue has "agranal" chloroplasts in which less than 1% of the thylakoids become appressed.

Chloroplast Replication

Two phases of plastid replication occur during leaf development and the plastid complement per cell increases at least three- or fourfold (Possingham, 1980). The first phase of plastid division occurs in the meristematic cells where proplastid and cellular divisions appear to be interdependent and the plastid number is maintained. Replication of plastids in developing guard cells has also been described in detail (Butterfass, 1979). The second phase of plastid division occurs when chloroplasts have grown to half their final size in cells which have ceased to divide or to synthesise nuclear DNA. The number of plastid division cycles is a major factor determining the chloroplast number in the mature leaf cell and can result in an increase as great as tenfold, e.g. in spinach, but more usually three or fourfold. Chloroplast division is rapid and generally completed in less than an hour (Leech et al., 1981), despite the morphological complexity of the dividing chloroplasts which have numerous grana containing several thylakoids.

The characteristics of the chloroplast division process are consistent with a physical initiation (Leech et al., 1981) but the nature of the specific mechanisms which initiate and terminate chloroplast division are unknown (for a review see Possingham, 1980). The chloroplasts of a cell divide semi-synchronously along their short axes into two equal daughter plastids and the grana and cpDNA appear to be partitioned more or less equally. Since the cellular complement of chloroplasts is such a key factor in the photosynthetic capacity of mature mesophyll cells, the control of chloroplast division will certainly be investigated in greater detail in the future. It has been suggested that chloroplast division during normal leaf development may be associated with polyploidy but natural polyploidisation is not a characteristic of leaf cellular development as it is in root cells. In Triticum (wheat) this possibility has been specifically investigated and natural polyploidy discounted experimentally as a factor which could influence normal chloroplast division (Ellis _et al_., 1983).

The mean chloroplast number per cell appears to be species specific and the control of chloroplast number predominantly nucleotypic. For example in wheat hybrids, formed by the introduction of an alien nucleus differing in ploidy from the cytoplasm, the chloroplast number reflects the ploidy level of the nucleus (Dean & Leech, 1982c). The tight packing of chloroplasts in the mature mesophyll cells of crop plants grown under non-limiting conditions is very consistently observed. Under nutrient stress or in conditions of suboptimal day length, cell size and chloroplast number both generally remain unchanged but the chloroplasts are reduced in size. Chloroplast replication is associated with DNA replication and cpDNA synthesis can occur independently of nDNA synthesis (Tymms _et al_., 1982, 1983). In the spinach (Scott & Possingham, 1983) and wheat (Boffey & Leech, 1982) cpDNA synthesis ceases before chloroplast replication begins. Prior to plastid division the number of plastome copies can rise to over 1000 per plastid in some species but is lower after chloroplast replication is complete. The cpDNA in young spinach leaves is 20% of the total cell DNA and 5000 plastome copies are present in each cell, _i.e_. about 200 per chloroplast (Scott & Possingham, 1983). In hexaploid wheat each mature chloroplast has _ca_.300 plastome copies, _i.e_. there are over 40,000 plastome copies per cell. At present the significance of this massive synthesis of plastid DNA ahead of the major phase of chloroplast synthetic function can only be a matter for conjecture but it would appear that chloroplast DNA amounts are unlikely to be a limiting factor in development.

THE SYNTHESIS OF MOLECULAR COMPONENTS

It is now clear that the complexity of interactions induced by endogenous and environmental factors during leaf development are only likely to be unravelled by gaining a deeper insight into the molecular interactions involved. All the indications are that nuclear control dominates since the chloroplast RNA polymerases appear to be nuclear encoded and so the level of chloroplast transcription is directly affected by the level of nuclear transcription. However, there is a vast excess of chloroplast coding capacity over nuclear genes and it has been estimated that there are only 10-12 copies of each nuclear gene which codes for a chloroplast protein (Berry-Lowe et al., 1983; Broglie et al., 1983). The chloroplast has multiple copies of chloroplast ribosomal RNA genes which may be needed to ensure the rapid and massive synthesis of chloroplast ribosomes which occurs prior to the major phase of protein synthesis during leaf development. Increases in both chloroplast and cytoplasmic ribosome complements occur early in mesophyll cell development and large increases in protein accumulation continue until just before maximum chloroplast size is reached. Chloroplast ribosomes (70S) are synthesised at a rate of ca.2 million per hour per mesophyll cell in wheat and the synthesis continues for 18 hours: during the corresponding period cytoplasmic ribosomes increase by only 60% (Dean & Leech, 1982a). By quantifying the changes in the components of cells of different developmental age it is possible to relate the changes in the visual appearance of developing cells and developing chloroplasts to changes in patterns of synthesis or accumulation of their component molecules. This can be done by examining cells of differing age within a single young grass leaf (Fig. 2). For example, chloroplast division and growth, chlorophyll biosynthesis and protein accumulation continue for a time after the period of maximum accumulation of ribosomal RNA. Chloroplast division ceases before the major leaf soluble protein RuBPCase accumulates at maximal rate.

Protein Synthesis

Synthesis of Ribulose Bisphosphate Carboxylase. The synthesis of the major soluble protein RuBPCase dominates protein synthesis during leaf development and the major membrane protein associated with chlorophylls a and b in the thylakoid, the light-harvesting chlorophyll a/b protein complex also accumulates rapidly (Fig. 3). In addition to

Fig. 2. Changes in total RNA, DNA and protein/cell during cell development in the first leaf of Triticum aestivum cv. Maris Dove. (from Dean & Leech, 1982a).

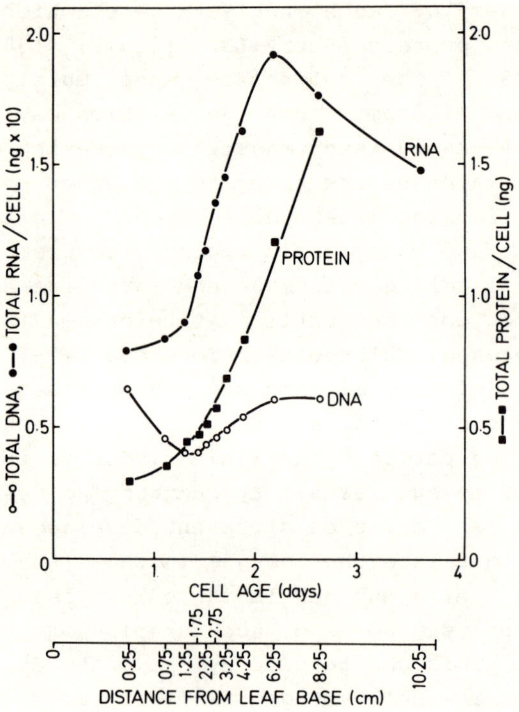

these two major components it has been estimated that chloroplasts contain at least 1000 other proteins. At present very little is known about the significance of the changes in these proteins, but quantitative and qualitative changes in their polypeptides occur during leaf differentiation. The polypeptides are separated, after reaction with detergents, using polyacrylamide gel electrophoresis (PAGE) and can be quantified by densitometric measurements and the incorporation of radioactive precursors detected on autoradiographs of the

Fig. 3. Densitometric scans of ^{35}S-methionine in vivo labelled polypeptides in young leaves of Triticum aestivum cv. Maris Dove. The positions of the large (L) and small (S) subunits of RuBPCase, the light harvesting complex (LHCP) and the α and β subunits of CF_1 are indicated. (Dean & Leech, unpublished).

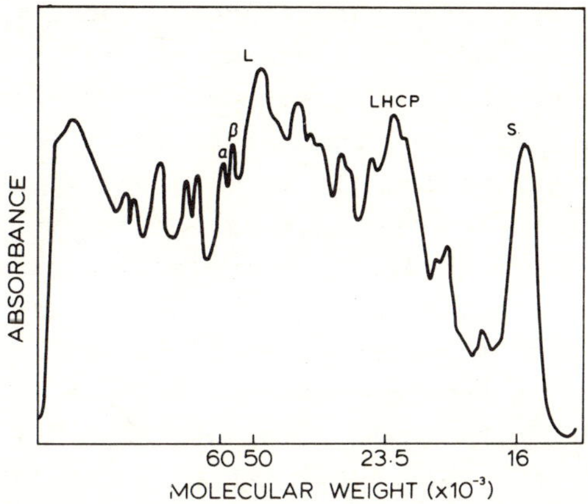

gels. Comparison of the spectra of leaf polypeptides with extracted polypeptides from the corresponding chloroplast, nuclear and cytoplasmic protein fractions allows identification of the location of the polypeptides in the leaf cell. For the chloroplast proteins, there is close coordination in the biosynthesis of the constituent subunits within individual proteins notably of RuBPCase and of the coupling factor. The individual proteins of the chloroplast are also synthesised at rates which are maintained rather constantly relative to each other throughout chloroplast development and replication. For example, the patterns in the early synthesis of the subunits of RuBPCase and the thylakoid proteins closely parallel each other. The patterns of in vivo synthesis of the leaf polypeptides revealed after feeding with ^{35}S-methionine are also closely related to the patterns of their accumulation. Close similarities are also found between the patterns of in vivo protein synthesis and in vitro translation using extracted mRNA and reticulocyte lysate preparations and suggest that the major control of this coordination occurs during transcription (Dean & Leech, unpublished). Indeed in those cases where the control of the synthesis of specific chloroplast proteins has been investigated in any detail, control primarily through the nuclear genes is almost always indicated.

Most of the recent detailed biochemical investigations of the control of the synthesis of leaf proteins have been of RuBPCase. RuBPCase is the most abundant protein in the chloroplast, in the plant and probably in the world. It is of great importance because of its role as the major carboxylating enzyme in photosynthesis and also has considerable interest because of its hybrid genetic characteristics. During wheat leaf development, the cellular content of RuBPCase increases twentyfold during three days of development and the rate of synthesis can approach 80 pg h^{-1} cell in very actively growing tissue. The amount of the protein which accumulates in a mature mesophyll cell is related quantitatively to nuclear genome size (Dean & Leech, 1982b) and in hexaploid wheat genetic lines a major control resides on nuclear chromosome 4A (Jellings et al., 1983). In the youngest leaf cells which have been analysed and whose plastids are small (less than 1 μm in diameter), the large (L) nuclear-encoded and small (S) chloroplast-encoded subunits of the RuBPCase holoenzyme are both present and the rates of the synthesis of the two subunits are closely synchronised (Nivison & Stocking, 1983; Dean & Leech, 1982c) during subsequent cellular development. The normally closely coordinated expression of chloroplast and nuclear genes in the synthesis of the L and S subunits respectively has been demonstrated in several species using different methods of analysis and reflects parallel rates of synthesis of the messenger RNAs (Dean & Leech, 1982c; Nelson et al., 1984). It is interesting that this coordinated synthesis of the two subunits still occurs in developing leaves of a genetic line of wheat in which an alien hexaploid nucleus has been introduced into diploid cytoplasm (Dean & Leech, 1982d). In many C4 plants, including *Zea mays*, the expression of the genes is regulated in a cell-specific manner: mesophyll cell chloroplasts lack RuBPCase while the enzyme is present in abundance in the chloroplasts of the bundle sheath cells (Hattersley *et al.*, 1977).

The chloroplast gene of the large subunit of RuBPCase has recently been cloned and sequenced from several species (see Dyer, 1984, for a review). The examination of the organisation of nuclear gene(s) encoding the small subunit precursor (pS) has shown that a family of genes encode for pS which in wheat show a rich array of genetic variation with the members of the gene family differing in both coding and non-coding regions (Broglie *et al.*, 1983). In petunia the nucleotide sequences of different pS genes show 9% divergence and distinct mRNAs are transcribed by these genes yet the amino acid sequence of the pS synthesised is highly conserved (Dunsmuir *et al.*, 1983a).

Synthesis of Thylakoid Proteins. During chloroplast differentiation, thylakoid membrane biogenesis is also very rapid and the proteins of the light-harvesting chlorophyll *a/b* protein complex (LHCP) are continuously inserted into the developing membrane. The light-harvesting chlorophyll a/b protein complex operates as a light-harvesting system and also in the transfer of excitation energy to PS1 and PS2 during photosynthesis. The physical structure of the complex is not well-understood but it contains nearly all the chlorophyll *b* and at least 50% of the chlorophyll *a* of the thylakoid. The chlorophylls are in non-covalent association with at least two but probably five different polypeptides and it is possible that different polypeptides associate with chlorophyll *a* and with chlorophyll *b* (Brown, 1983). The two major proteins, which are 24-28,000 in molecular weight (M_r) are synthesised as soluble precursors on free cytoplasmic ribosomes and are also nuclear-encoded. In the dark, in the absence of chlorophyll biosynthesis, the proteins of the complex disintegrate (Bennett, 1981). In developing barley and wheat leaves the accumulation of the complex closely follows the accumulation of pS and L of RuBPCase. In barley (Viro & Kloppstech, 1980) the synthesis of mRNA for the light-harvesting chlorophyll a/b complex precursors declines earlier than the synthesis of mRNA for pS or L. Many nuclear coding sequences for the major LHCP protein (Cab) have been recognised in petunia and five have been shown to be transcribed. The cDNA clones for Cab are characterised by more than 90% sequence conservation in the amino acid coding regions and extreme sequence divergence in the untranslated regions (Dunsmuir et al., 1983b). The mRNAs are distinct and give rise to a set of distinguishable Cab polypeptides. The physiological significance of this rich array of genetic variation is as yet unknown but it is possible that different genes may be expressed at different stages in the life cycle or in response to changing environmental stimuli and this exciting possibility will certainly be actively investigated.

The available information about changes in the other soluble and thylakoid polypeptides during leaf development is confined to descriptions of their relative rates of accumulation. Chloroplast coupling factor (CF_1) is a multisubunit protein and the synthesis of its α, β and γ subunits is coordinated. The accumulation of CF_1 in the plastid peaks early during the period of thylakoid biogenesis, before extensive appression and granal assembly begins, and while RuBPCase is till being rapidly synthesised. The pattern of synthesis of CF_1 is in line with evidence which suggests it is confined to the exposed thylakoid membranes. Cytochrome f is probably present in both appressed and exposed thylakoid membranes, so it is particularly interesting that its synthesis continues longer than the synthesis of either CF_1 or RuBPCase.

Light is a major external factor which modulates leaf development and chloroplast development. At the cellular level the analysis of light effects will be greatly assisted in the future by the application of the techniques of molecular genetics. In a recent paper on developing maize leaves the effects of light on the synthesis of mRNAs of the four major chloroplast proteins was examined following this approach and the proteins examined were the L and S subunits of RuBPCase, LHCP and phosphoenolpyruvate carboxylase (PEPcase) (Nelson *et al.*, 1984). Their mRNAs levels were measured by hybridisation with cloned DNA probes and the results suggest that the initial synthesis of mRNAs of all four proteins is light-independent but that light influences the level to which the mRNAs accumulate. Protein accumulation was measured by immunodetection and it was shown that LHCP synthesis also required light. The LS subunit synthesis and the synthesis of PEP carboxylase was light-independent and the synthesis of these two proteins closely followed mRNA accumulation.

In greening etiolated plants chlorophyll biosynthesis, initiated at the outset of illumination, is the trigger for the stepwise accumulation of the macromolecular complexes utilised in thylakoid biogenesis. In normal leaves the membranes of the youngest plastids contain chlorophyll *a* and *b* and biosynthesis continues throughout development (Leech & Baker, 1983). The dependence of the continuing accumulation of LHCP on light may well be related to the syntheses of chlorophylls *a* and *b*.

Synthesis of Chloroplast Lipids

The properties of the thylakoid membrane reflect not only the functions of the constituent protein-chlorophyll complexes but also the properties of the numerous components of the membranes where lipids dominate quantitatively. Because of the role of lipids in maintaining membrane stability and creating the appropriate environment for photosynthetic function, the biogenesis of chloroplast lipids has been studied in detail. Their assembly in the thylakoid membranes produces a membrane of unique composition. Leaf chloroplast thylakoids are 50% lipid by weight and current models envisage the thylakoid membrane as a fluid mosaic with the protein complexes embedded in the lipid matrix. Lipid-protein and lipid-lipid interactions ensure both membrane stability and also the efficient operation of photosynthesis over the wide temperature range experienced by leaves. The specificity of these interactions depends on the molecular properties of individual thylakoid lipids and in view of its highly specialised function, it is perhaps not surprising that the photosynthetic membrane has a unique acyl lipid complement in which highly polyunsaturated glycerol lipids are the main components.

The synthesis of lipids during chloroplast membrane biogenesis has been followed in particular detail with a view to the identification of both the specific lipid molecular species required for the correct assembly of the thylakoid membrane and also the specific molecules or assemblies of lipids needed for efficient development of light absorption and energy transduction functions. The major chloroplast thylakoid lipids, accounting for ca.75% of the membrane lipids, are the neutral galactolipids, monogalactosyldiacylglycerol (MGDG) and digalactosyldiacyl glycerol (DGDG). Two additional neutral lipids sulfoquinovosyldiacylglycerol (SQDQ) and phosphatidyl glycerol (PG) each contribute ca.10% to the photosynthetic membrane. The acyl moieties of the neutral galactolipids are almost exclusively α-linolenoyl residues (18:3) or 16:3 in some plants. During chloroplast development in grass leaves very large increases per plastid occur in all the thylakoid lipids which become progressively more polyunsaturated (for a review see Leech & Walton, 1983). The largest increases are a sixfold increase in MDGD and a fourfold increase in DGDG. The increases in PG are particularly interesting since at all stages in plastid development the increase in PG parallels the increase in chlorophyll and at maturity one molecule of PG is present for every three chlorophyll molecules. This stoichiometric relationship tempted the speculation that PG is associated specifically with the chlorophyll-protein complexes in the thylakoid membrane - a suggestion supported by the discovery that the only major qualitative change in lipid molecular species, which occurs during plastid development, is the appearance of a new molecular species of PG containing Δ^3-trans hexadecanyl

$(16{:}1)_{t\text{-}3}$ residues. It has also often been suggested that a high content of 18:3 or 16:3 residues (α-linolenoyl) is an essential requirement for the rapid and extensive appression of membranes into grana during chloroplast development, but recent experimental observations contradict this contention (Leech & Walton, 1983).

CONCLUSIONS

The analyses of changes in cellular components during leaf development have been predominantly descriptive and mainly relate to changes in mesophyll chloroplasts. Observations of other cellular components and other leaf tissues are necessary for a more comprehensive understanding of the integration of leaf development. The dependence of plastid differentiation on the continuing integrated function of the developing mesophyll cell is not in doubt and the sequential changes during chloroplast development and replication have now been described in some detail. The youngest proplastids with limited bithylakoid development are already photochemically competent. Chloroplast development is accompanied by large increases in the major plastid components and quantitative rather than qualitative changes predominate: smaller changes in quantitatively minor components, which might be significant, have yet to be extensively investigated. Moreover, the mechanisms, which ensure the continued integrated assembly of the components during plastid growth, are largely unknown. The next phase in investigative research will certainly include programmes designed to obtain more detailed knowledge of the controls which operate both in time and space, during different phases of leaf development and in adjacent cells and tissues of the same leaf. Techniques, which allow the non-destructive monitoring, e.g. *in vivo* spectroscopic measurements, of developmental changes over extended time periods and/or over short distances intracellularly or intercellularly, e.g. immunocytochemical methods, will be of particular value. Another aim in future work should be the devising of methods for measuring the continuum of development to complement the episodic sampling methods currently in use. Certainly cell biologists interested in leaf development will benefit increasingly from progress in molecular biology, which can be expected to provide both the inspirational and analytical tools necessary for their studies.

REFERENCES

Baker, N.R. & Leech, R.M. (1977). Development of photosystem I and photosystem II activities in leaves of light-grown maize (Zea mays). Plant. Physiol., 60, 640-4.

Baker, N.R., Markwell, J.P.,Bradbury, M., Baker, M.G. & Thornber, J.P. (1983). Thylakoid protein kinase activity and associated control of excitation energy distribution during chloroplast biogenesis in wheat. Planta, 159, 151-8.

Baker, N.R., Webber, A.N., Bradbury, M., Markwell, J.P., Baker, M.G. & Thornber, J.P. (1984). Development of photochemical competence during growth of the wheat leaf. In Biosynthesis of the Photosynthetic Apparatus: Molecular Biology, Development and Regulation, eds. J.P. Thornber, L.A. Staehelin & R.B. Hallick, pp. 237-55. New York: Alan R. Liss Inc.

Bennett, J. (1981). Biosynthesis of the light-harvesting chlorophyll a/b protein. Polypeptide turnover in darkness. Eur. J. Biochem., 118, 61-70.

Berry-Lowe, S.L., McKnight, T.D., Shah, D.M. & Meagher, R.B. (1982). The nucleotide sequence, expression and evolution of one member of a multigene family encoding the small subunit of ribulose-1,5-bisphosphate carboxylase in soybean. J. Mol. Appl. Genet., 1, 483-98.

Boffey, S.A., Ellis, J.R., Sellden, G. & Leech, R.M. (1979). Chloroplast division and DNA synthesis in light-grown wheat seedlings. Plant Physiol., 64, 502-5.

Boffey, S.A. Sellden, G. & Leech, R.M. (1980). The influence of cell age on chlorophyll formation in light-grown and etiolated wheat seedlings. Plant Physiol., 65, 680-4.

Boffey, S.A. & Leech, R.M. (1982). Chloroplast DNA levels and the control of chloroplast division in light-grown wheat leaves. Plant Physiol., 69, 1387-91.

Brangeon, B. & Mustardy, L. (1979). The ontogenic assembly of intra-chloroplastic lamellae viewed in 3-dimension. Biol. Cellul., 36, 71-80.

Broglie, R., Coruzzi, G., Lamppa, G., Keith, B. & Chua, N.H. (1983). Structural analysis of nuclear genes coding for the precursor to the small subunit of wheat ribulose-1,5-bisphosphate carboxylase. Biotechnology, 1, 55-61.

Brown, J.S. (1983). A new evaluation of chlorophyll absorption in photosynthetic membranes. Photosyn. Res., 4, 375-83.

Butterfass, T.H. (1979). Patterns of chloroplast reproduction. Cell Biology Monographs Vol. VI. New York: Springer.

Butterfass, T.H. (1983). A nucleotypic control of chloroplast reproduction. Protoplasma, 118, 71-4.

Chapman, D.J. & Leech, R.M. (1979). Changes in pool sizes of free amino acids and amides in leaves and plastids of *Zea mays* during leaf development. Plant Physiol., 63, 567-72.

Dale, J.E. (1982). The growth of leaves. London: Edward Arnold.

Dean, C. & Leech, R.M. (1982a). Genome expression during normal leaf development. 1. Cellular and chloroplast numbers and DNA, RNA and protein levels in tissues of different ages within a seven-day old wheat leaf. Plant Physiol., 69, 904-910.

Dean, C. & Leech, R.M. (1982b). Genome expression during normal leaf development. 2. Direct correlation between ribulose bisphosphate carboxylase content and nuclear ploidy in a polyploid series of wheat. Plant Physiol., 70, 1605-8.

Dean, C. & Leech, R.M. (1982c). The coordinated synthesis of the large and small subunits of ribulose bisphosphate carboxylase during early cellular development within a seven day wheat leaf. FEBS Lett., 140, 113-6.

Dean, C. & Leech, R.M. (1982d). The coordinated synthesis of the subunits of ribulose bisphosphate carboxylase in a wheat line with alien cytoplasm. FEBS Lett., 144, 154-6.

Dunsmuir, P., Smith, S. & Bedbrook, J. (1983a). A number of different nuclear genes for the small subunit of RuBPCase are transcribed in *Petunia*. Nuc. Acid Res., 11, 4177-83.

Dunsmuir, P., Smith, S.M. & Bedbrook, J. (1983b). The major chlorophyll a/b binding protein of *Petunia* is composed of several polypeptides encoded by a number of distinct nuclear genes. J. Mol. Appl. Genet., 2, 285-300.

Dyer, T.A. (1984). The chloroplast genome - its nature and role in development. *In* Chloroplast Biogenesis, Topics in Photosynthesis, Vol. 5, eds. N.R. Baker & J. Barber, in press. Amsterdam: Elsevier.

Ellis, J.R., Jellings, A.J. & Leech, R.M. (1983). Nuclear DNA content and the control of chloroplast replication in wheat leaves. Planta, 157, 376-80.

Hattersley, P.W., Watson, L. & Osmond, C.B. (1977).

In situ immunofluorescent labelling of ribulose-1,5-bisphosphate carboxylase in leaves of C3 and C4 plants. Aust. J. Physiol., 4, 523-39.

Hawke, J.C., Rumsby, M.G. & Leech, R.M. (1974). Lipid biosynthesis in green leaves of developing maize. Plant Physiol., 53, 551-61.

Jellings, A.J. & Leech, R.M. (1982). The importance of quantitative anatomy in the interpretation of whole leaf biochemistry in species of Triticum, Hordeum and Avena. New Phytol., 92, 39-48.

Jellings, A.J., Leese, B.M. & Leech, R.M. (1983). Location of chromosomal control of ribulose bisphophate carboxylase amounts in wheat. Mol. Gen. Genet., 192, 272-4.

Leech, R.M., Thomson, W.W. & Platt-Aloia, K.A. (1981). Observations on the mechanism of chloroplast division in higher plants. New Phytol., 87, 1-9.

Leech, R.M. & Baker, N.R. (1983). The development of photosynthetic capacity in leaves. In Growth and Functioning of Leaves, eds. J. Dale & F. Milthorpe, pp. 271-307. Cambridge: Cambridge University Press.

Leech, R.M. & Walton, C.A. (1983). Modification of fatty acid composition during chloroplast ontogeny and the effects on thylakoid appression and primary photochemistry. In Biosynthesis and Function of Plant Lipids, eds. W.W. Thomson, J.B. Mudd & M. Gibbs, pp. 56-80. Proceedings of the 6th Annual Symposium in Botany, Rockville, Maryland: American Society of Plant Physiologists.

Maksymowych, R. (1973). Analysis of Leaf Development. Cambridge: Cambridge Univeristy Press.

Maksymowych, R. & Kettrick, M.A. (1970). DNA synthesis, cell division and cell differentiation during leaf development of Xanthium pennslyvanicum. Am. J. Bot., 57, 844-9.

Miranda, V., Baker, N.R. & Long, S.P. (1981a). Anatomical variation along the length of the Zea mays leaf in relation to photosynthesis. New Phytol., 88, 595-605.

Miranda, V., Baker, N.R. & Long, S.P. (1981b). Limitations of photosynthesis in different regions of the Zea mays leaf. New Phytol., 89, 179-90.

Nelson, T., Harpster, M.H., Mayfield, S.P. & Taylor, W.C. (1984). Light-regulated gene expression during maize leaf development. J. Cell. Biol., 98, 558-64.

Nivison, H.T. & Stocking, C.R. (1983). Ribulose bisphosphate carboxylase synthesis in barley leaves. A developmental approach to the question of coordinated subunit synthesis. Plant Physiol., 73, 906-11.

Possingham, J.V. (1980). Plastid replication and development in the life cycle of higher plants. Ann. Rev. Plant

Physiol., 31, 113-29.

Scott, N.S. & Possingham, J.V. (1983). Changes in chloroplast DNA levels during growth of spinach leaves. J. Exp. Bot., 34, 1756-67.

Taylor, J.A. & Mackender, R.O. (1977). Plastid development in the first leaf of *Avena sativa*. Plant Physiol., 59, S 10.

Tymms, M.J., Scott, N.S. & Possingham, J.V. (1982). Chloroplast and nuclear DNA content of cultured spinach leaf discs. J. Exp. Bot., 33, 831-7.

Tymms, M.J., Scott, N.S. & Possingham, J.V. (1983). DNA content of *Beta vulgaris* chloroplasts during leaf cell expansion. Plant Physiol., 71, 785-8.

Viro, M. & Kloppstech, K. (1980). Differential expression of the genes for ribulose-1,5-bisphosphate carboxylase and light-harvesting chlorophyll a/b protein in the developing barley leaf. Planta, 150, 41-5.

Whatley, J.M. (1977). Variations in the basic pathway of chloroplast development. New Phytol., 78, 407-20.

Whatley, J.M. (1979). Plastid development in the primary leaf of *Phaseolus vulgaris*: variations between different types of cell. New Phytol., 82, 1-10.

ENERGY TRANSDUCTION DURING LEAF GROWTH

Neil R. Baker
Department of Biology, University of Essex,
Colchester, CO4 3SQ, United Kingdom.

INTRODUCTION

A major function of the mature leaf of an autotrophic higher plant is to transduce light energy into chemical energy, which is then utilized for the synthesis of carbohydrates from carbon dioxide. The assimilation products of this photosynthetic activity can then be translocated to and catabolised by other organs to provide energy for growth and other physiological activities. Photosynthesis involves a highly complex series of physico-chemical reactions which are dependent upon many specific structural characteristics and enzyme activities within the leaf cells. At early stages of leaf growth the components of the photosynthetic apparatus have to be synthesised and assembled into functional units in order to allow the leaf to utilize light energy for carbon assimilation and achieve an autotrophic mode of existence. During the development of a fully functional photosynthetic apparatus a leaf is obviously dependent upon a supply of chemical energy from sites such as the seed, cotyledons and older leaves. Conversion of the energy contained within the translocated organic compounds received by the developing leaf into ATP and reducing power, which are required to drive the biosynthetic reactions, is ultimately dependent upon the respiratory activity of the leaf. Respiration being defined as the oxidation of carbohydrates via glycolysis and the tricarboxylic acid cycle or via the oxidative pentose phosphate pathway; oxidation of the products of these processes by molecular oxygen occurs via the mitochondrial electron transport system.

A knowledge of the respiratory and photosynthetic activities of a developing leaf is insufficient to provide a complete understanding of energy metabolism occurring during growth since the process of carbon assimilation in photosynthetic tissues is often intimately associated with photorespiratory activity, which itself is intimately associated with nitrogen metabolism. Photorespiration is the carbon dioxide evolution and oxygen consumption that occurs during light-induced carbon dioxide assimilation from sources other than the tricarboxylic acid cycle. Although it is evident that many regulatory interactions occur between the respiratory, photorespiratory and photosynthetic metabolic processes in the mature leaf, our understanding of these interactions is still in its infancy and consequently our knowledge of the interactive associations between these energy transducing processes during leaf growth is minimal. The developing leaf, in which there is differential development of the respiratory, photosynthetic and photorespiratory activities, would appear to offer a potentially useful system with which to examine the nature of interactions between these processes. In this chapter the limited information available on respiratory, photosynthetic and photorespiratory activities during leaf growth is briefly reviewed and consideration is given to the possible interactions that may occur between these processes as they develop.

RESPIRATION

Although there have been few detailed studies of respiratory activities during leaf development, it is generally considered that there is an initial increase in the dark respiration rate of a leaf at early stages of expansion, which is then followed by a gradual decline in activity. If the respiration rate is expressed on the basis of unit leaf area, rather than per leaf, similar trends are also observed. The initial increase in respiration presumably reflects mitochondrial development in the leaf cells, whilst the decline is attributed to the increase in the photosynthetic ability of the leaf, which results in a decreased requirement for respiratory activity. Since the maximum rate of oxidative phosphorylation per unit protein of isolated mitochondria increases during the four hours after the onset of irradiation of etiolated Avena leaves (Wellburn, 1982), it would appear that development of mitochondria in leaf tissue has a specific light requirement, which is probably mediated by phytochrome (Hampp & Wellburn, 1982). However, the role of light in determining respiratory activity in developing leaves is unclear because it is often claimed that mitochondrial respiratory activities in mature leaves are significantly inhibited by light. Estimates of the proportion of the dark respiration that is inhibited by light have been made during ontogeny (Peisker, Ticha & Catsky, 1981) and have suggested that large fluctuations in the ability of light to inhibit respiratory activity occur during leaf development. The high cytosolic ATP/ADP ratio that is generated by photosynthetic activity of the thylakoids when a leaf is exposed to light is thought to inhibit mitochondrial respiratory activity. However, recent evidence suggests that if mitochondrial respiration is suppressed in the light, and it may not be, the inhibition is not mediated by an increase in the adenylate energy charge of the cytoplasm (Kelly, 1983).

The fact that a leaf contains populations of both autotrophic and heterotrophic cells increases the complexity of interpreting the significance of changes in respiration rate during development. Although in a mature leaf the majority of cells are autotrophic and may have little respiratory activity, a significant proportion of cells are heterotrophic and can be expected to have high respiration rates in order to maintain their metabolic activities. At early stages of leaf growth, before the development of functional photosynthetic systems, cells will be heterotrophic, although, of course, many cells will have the potential to become autotrophic. Cells will switch from the heterotrophic to an autotrophic mode of existence as chloroplast development proceeds and photosynthetic activity becomes sufficiently large to satisfy the cells' metabolic demands.

PHOTOSYNTHESIS

Many studies on a wide range of plant species have shown that the capacity for light-induced carbon assimilation increases with leaf age. The complex series of reactions that comprise the process of photosynthesis can be conveniently divided into two sections to facilitate consideration of the development of photosynthetic activities in the leaf. Firstly, the processes involved with light trapping, electron transport and ATP synthesis, which are all associated with the thylakoid membranes of the chloroplasts, will be examined. Consideration will then be given to the biochemical reactions involved with carbon assimilation.

Photochemical Reactions

The pigment systems in the leaf that are responsible for the harvesting and transduction of light energy into chemical energy are located on the thylakoid membranes. The pigments, chlorophylls and carotenoids, are bound to membrane proteins, and these pigment-protein complexes can be divided into three categories on the basis of their function. These are (i) a photosystem 1 (PS1) complex containing about 30% of the total chlorophyll content of the leaf and responsible for the transfer of trapped light energy to the PS1 reaction centre, P700; (ii) a photosystem 2 (PS2) complex containing up to 20% of the total chlorophyll and responsible for the transfer of light energy to the reaction centre of PS2, P680; (iii) a light-harvesting chlorophyll *a/b* protein complex (LHCP) comprising 40-60% of the total chlorophyll and which is generally thought to be closely associated with and capable of efficient excitation energy transfer to PS2 complexes, however it can also transfer energy to PS1 complexes (Thornber *et al*., 1979; Barber, 1983). The reaction centres of PS1 and PS2, when excited by light energy, can transfer electrons to components of the electron transport system and drive the overall transfer of electrons from water to NADP. The coupling of the pigment-protein complexes to the electron transport system is shown in Fig. 1. Under certain conditions oxygen, rather than NADP, may act as the terminal electron acceptor with electrons being passed from ferredoxin to oxygen in a reaction often termed the Mehler reaction (see Fig. 1). Reduced ferredoxin is also capable of transferring electrons to oxidised plastoquinone, this process results in a cyclic electron flow driven by PS1 (Barber, 1983). As water is photo-oxidised by PS2 and electrons are transferred through plastoquinone to PS1, protons are released into the intrathylakoid space and consequently a proton electro-chemical gradient is generated across the thylakoid membrane. The potential energy of this proton electro-chemical gradient can be utilised to produce ATP via a membrane bound ATPase, termed the coupling factor; protons are translocated from the intrathylakoid space to the stroma via the coupling factor and stromal ADP is phoshorylated (Hinckle & McCarty, 1978).

Studies with greening etiolated leaves have demonstrated the sequence of development of light-harvesting end electron transport apparatus (Baker, 1984). Shortly after the onset of chlorophyll synthesis, which occurs when etiolated leaves are exposed to light, photochemically competent PS1 and PS2 complexes are observed, but LHCP is absent. However, the ability to transfer electrons between PS2 and PS1 and perform non-cyclic electron transport from water to NADP develops later. With the development of non-cyclic electron transport proton pumping across the thylakoid membranes occurs, but at early stages of greening these membranes appear to be leaky to protons with the result that proton gradients cannot be maintained across the membranes. The efficient return of protons back across the membranes does not occur via the coupling factor and thus the potential energy gained in the light-induced pumping of protons into the intrathylakoid space cannot be utilised for ATP synthesis at these early stages of development. The

Fig. 1. Light-harvesting apparatus and photosynthetic electron transport system of the thylakoid membrane. Open arrows represent transfer of light energy, solid arrows represent the reactions of the electron transport sequence. LHCP, light-harvesting chlorophyll a/b protein complexes; PS2, photosystem 2; PS1, photosystem 1; P680, reaction centre of PS2; Q, primary electron acceptor of PS2; PQ, plastoquinone; PQH_2, reduced plastoquinone; cyt f, cytochrome f; PC, plastocyanin; P700, reaction centre of PS1; Fd, ferredoxin.

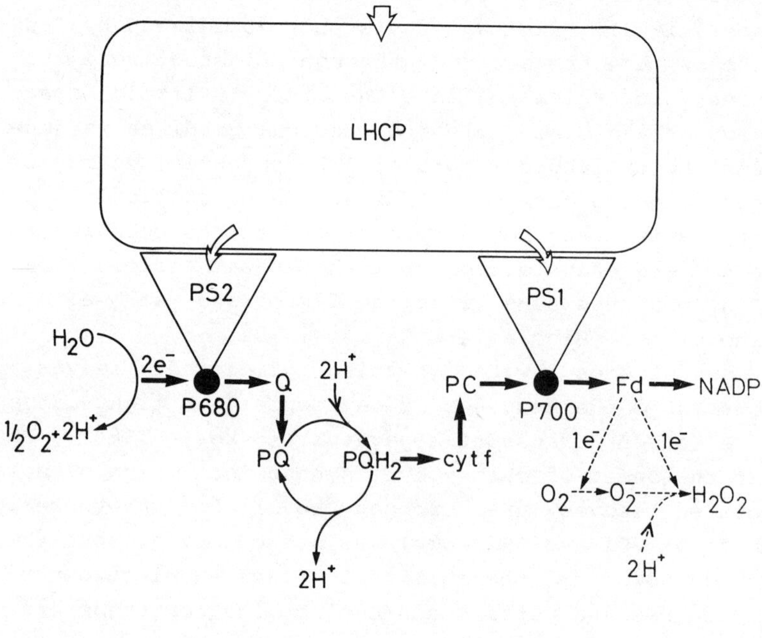

ability to maintain a proton gradient across the thylakoid membranes rapidly develops and the membranes then become capable of transducing light energy into both NADPH and ATP. The accumulation of LHCP by the membranes as greening of etiolated leaves proceeds increases the number of chlorophyll molecules energetically associated with a single reaction centre and results in an increase in the efficiency of the transduction of light energy into chemical energy. For more detailed considerations of the biogenesis of the photochemical apparatus the reader is referred to recent reviews by Leech & Baker (1983) and Baker (1984).

It is somewhat unfortunate that the majority of studies of the development of photochemical activities have been made on greening etiolated leaves because the pattern of plastid biogenesis in these leaves is rather different to that occurring in leaves growing under natural diurnal light regimes (Leech & Baker, 1983). In general, as chlorophyll accumulates in the greening etiolated leaf there is a gradual increase in the efficiency of harvesting light energy for photosynthetic reactions, this being demonstrated by a decrease in the light intensity required to saturate the photosynthetic activities. However, recent studies of PS2 photochemical activities during chloroplast development in wheat leaves growing under a diurnal light regime have suggested that this is not the case in light-grown leaves since the efficiency of light conversion into PS2 photochemistry appears to similar throughout chloroplast biogenesis (Webber _et al_., 1984). This difference between greening dark- and light-grown leaves would seem to be attributable to differences in the pattern of accumulation of LHCP into the developing thylakoids. In greening etiolated leaves the ratio of chlorophyll _a_: chlorophyll _b_ is very high initially and then decreases with time, indicating that LHCP is present in negligible amounts at early stages of greening but progressively increases as a proportion of the total chlorophyll content of the membranes, until a constant level is reached similar to that in the mature chloroplast. However, in light-grown leaves the chlorophyll _a_: chlorophyll _b_ ratio may not change significantly (Baker & Leech, 1977; Webber _et al_., 1984); ratios similar to those in the mature leaf are found at the earliest stages of leaf development, implying that the ratio of LHCP to PS1 and PS2 complexes is remaining constant throughout development. Clearly an increase in the ratio of LHCP to PS1 and PS2 complexes during leaf growth would make the photochemical apparatus potentially more efficient at harvesting and transducing light energy into chemical energy. It should be emphasised that although the actual mechanisms by which the components of the photochemical apparatus are synthesised and organised into physiologically functional units appears to be similar in both greening etiolated and light-grown leaves, the relative rates of accumulation of specific components of the apparatus can be considerably different. Consequently, care should be taken when attempting to extrapolate information from greening etiolated leaves to predict the situation in light-grown leaves.

Light-grown wheat leaf tissue at the earliest measurable stage of development has been found to exhibit light-induced oxygen evolution (Webber et al., 1984), indicating the presence of a functional non-cyclic electron transport system. However, recent studies on the thylakoids isolated from the developing wheat leaf have shown that although an efficient light-harvesting and electron transport is present at the very early stages of leaf development, the thylakoid membranes do not appear to be able to maintain a proton gradient and are therefore unable to produce ATP (Webber, A.N. & Baker, N.R., unpublished). However, this inability to maintain a proton gradient at early stages of development would not appear to be of great significance to the energetics of leaf growth since the ability to photochemically generate ATP soon appears during the lag phase of leaf growth. The presence of photochemically competent thylakoids at early phases of leaf growth ensures that the developing chloroplast can potentially generate ATP and reducing power for biosynthetic processes. However, the capacity for the generation of such chemical energy will be determined by the thylakoid membrane complement of the developing tissue. The chlorophyll content of the tissue would appear to be a reasonable indicator of the thylakoid membrane complement in the wheat leaf. At early stages of leaf growth, the capacity for light energy transduction will be rather insignificant compared to the potential supply of chemical energy from respiratory processes. This situation changes rapidly as chlorophyll accumulates in the growing leaf.

It is well established that the photochemical apparatus of thylakoids can be severely damaged by light if the NADPH and ATP generated by electron transport are not rapidly utilised thus maintaining a supply of oxidised terminal electron acceptors and ADP. Such photoinhibition is associated with PS2 photochemical activities and has profound implications for the energetics of leaf photosynthesis (Osmond, 1981). At early stages of chloroplast development the rate of reduction of the PS2 primary electron acceptor, Q, is considerably greater than the rate of utilisation of reducing power produced, thus the potential for photoinhibition of PS2 photochemical activities is great. Preliminary studies (Bradbury, M. & Baker, N.R., unpublished) have indicated that the developing photosynthetic membranes are more susceptible to photodamage than mature thylakoids, yet clearly massive photodamage of the developing thylakoids does not occur _in vivo_. It would appear that the developing chloroplasts have a mechanism for successfully dissipating trapped light energy and thereby preventing significant amounts of photoinhibition occurring. The mature chloroplast is capable of regulating the distribution of light absorbed by the LHCP between PS1 and PS2 by a mechanism known as the State 1- State 2 transition, which is thought to be controlled by phosphorylation of the LHCP (Barber, 1982; Haworth _et al_., 1982). Such a mechanism allows the rates of non-cyclic and cyclic electron transport to be controlled and thus provides a regulation of the ratio NADPH:ATP produced by the thylakoids. The State transition may also be important in the prevention of photoinhibitory damage to PS2, especially at early phases of leaf growth. In the wheat leaf the ability to perform a State transition is not observed at the earliest stages of thylakoid biogenesis, however this ability rapidly develops (Webber _et al_., 1984), but it is not clear whether the State transition actually plays a major role in the protection of PS2 from photodamage in the developing plastid. An attractive alternative protective mechanism, but as yet also unproven, is the possibility that developing thylakoids may be able to efficiently transfer electrons directly to oxygen. If electrons were transferred to oxygen via ferredoxin (Fig. 1) this would allow non-cyclic photophosphorylation to occur without a concurrent production of NADPH. Alternatively oxygen may be able to accept electrons directly from PS2; this being the case, light energy transduction would result in oxygen reduction with no potentially metabolically useful reducing power or ATP being produced.

Photosynthetic Carbon Assimilation

Light-dependent carbon dioxide assimilation in the mature leaf is dependent upon a supply of carbon dioxide to the carboxylation sites within the leaf, the presence of the enzymes necessary to perform the carboxylation and a supply of ATP and NADPH to drive the reactions and reduce the carbon dioxide. The rate of carbon assimilation may be limited by the rate of diffusion of carbon dioxide from the atmosphere to the carboxylation sites, the concentration of active enzymes or the rate of supply of ATP and/or NADPH. The resistance to carbon dioxide assimilation within a leaf, r_L, is conveniently divided into three major components; the resistance to carbon dioxide exchange across the boundary layer at the leaf's surface, r_a, the resistance of carbon dioxide movement through the stomata, r_s, and the resistance to carbon dioxide assimilation in the mesophyll liquid phase, r_m, which has both transport and carboxylation components. The magnitudes of r_m and r_s are considerably greater than that of r_a during leaf development (Catsky *et al.*, 1976; Miranda *et al.*, 1981). Decreases in both r_m and r_s do occur during leaf expansion. At early stages of development r_s is small compared to r_L, however as expansion proceeds r_s becomes an increasingly important component of r_L (Catsky *et al.*, 1976; Miranda *et al.*, 1981).

The supply of ATP and reducing power to the carboxylation sites may be a potential determinant of the magnitude of r_m during development, however the high rates of light-induced oxygen evolution relative to those for carbon dioxide fixation often observed at early stages of leaf development (e.g. Miranda et al., 1981) would suggest that this is unlikely. No comprehensive studies of the development of all of the enzymes of the photosynthetic carbon reduction cycle have been made, but it has been demonstrated frequently that the carboxylation activity per unit leaf area of ribulose 1,5-bisphosphate carboxylase-oxygenase (RuBPc-o) increases during leaf development (e.g. Baker & Hardwick, 1973; Callow, 1974). Since the specific activity of this enzyme activity was found to be relatively constant throughout development, the increase in activity per unit leaf area can be attributed to an increase in the concentration of the enzyme during leaf growth. The activities of a number of key enzymes associated with the C4 pathway of carbon assimilation in *Zea mays* have also been shown to show a marked increase during leaf development (Williams & Kennedy, 1978; Miranda et al., 1981). It is hoped that more effort will be put into the study of the carbon assimilatory pathways and their constituent enzyme activities during leaf development in the future. This is essential if a good understanding of the potential sites of limitation to carbon assimilation during leaf growth is to be achieved.

PHOTORESPIRATION

RuBPc-o can also oxidise ribulose 1,5-biphosphate (RuBP) to phosphoglycolate at physiological carbon dioxide concentrations and consequently both oxygen and carbon dioxide compete as substrates for the active site of the enzyme (Lorimer, 1981). Phosphoglycolate produced in the chloroplast is exported and converted to glycolate by peroxisomes. Recovery of the majority of the glycolate carbon for future potential use in carbon assimilatory pathways is achieved by the integration of a series of reactions occurring in the chloroplasts, peroxisomes and mitochondria. These reactions, which evolve carbon dioxide and consume oxygen, are often collectively termed the photorespiratory carbon oxidative pathway and are shown in

Fig. 2 (Osmond, 1981). The recovery of glycolate carbon by this photorespiratory pathway is energetically expensive and consequently reduces the quantum yield of photosynthetic carbon assimilation (Osmond, 1981).

Fig. 2. Reaction sequences of the photosynthetic carbon reduction cycle and photorespiratory carbon oxidation pathway. The compartmentation of the reactions between the chloroplasts, peroxisomes and mitochondria is shown. Triose-P, triose phosphate; FbP, fructose 1,6-bisphosphate; F6P, fructose 6-phosphate; E4P, erythrose 4-phosphate; SbP, sedoheptulose 1,7-bisphosphate; S7P, sedoheptulose 7-phosphate; RuP, ribulose 5-phosphate; RubP, ribulose 1,5-bisphosphate; 3-PGA, 3-phosphoglycerate; P-glycolate, phosphoglycolate. Modified from Osmond (1981).

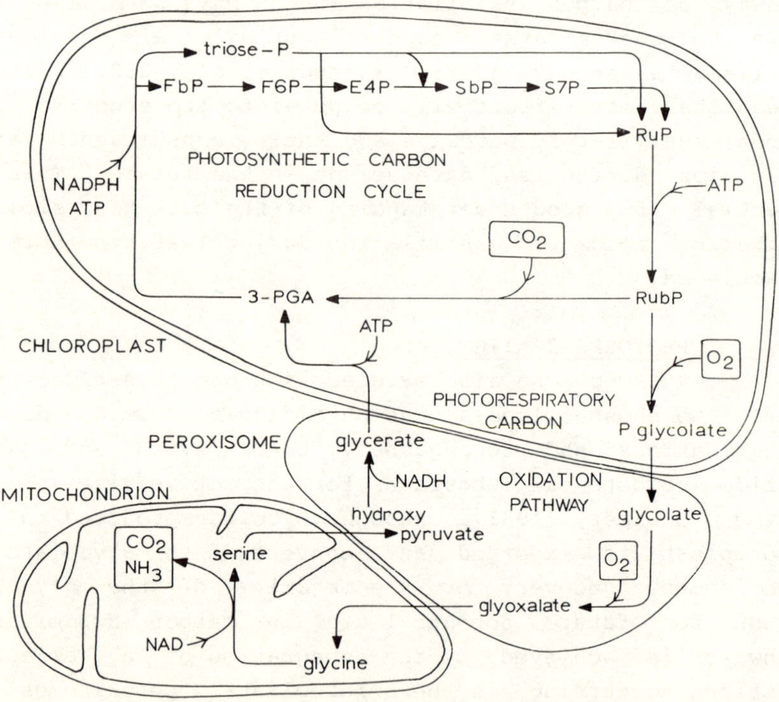

Unfortunately the photorespiratory activities of developing leaves have received relatively little attention to date, presumably because of the practical difficulties in accurately determining photorespiratory rate. The majority of studies that have been made have concluded that photorespiratory activity decreases during leaf ontogeny. In contrast a number of reports have suggested that an increase in photorespiration rate occurs during development, however doubts have been raised about the validity of this conclusion (Kennedy & Johnson, 1981). As is the case with the photosythetic carbon reduction cycle, it is to be hoped that considerably more research will be carried out in the future in order to determine the changes in photorespiratory activity during leaf ontogeny since a knowledge of these processes is essential to achieve a comprehensive understanding of the energetics of leaf growth.

Besides photorespiration acting as a mechanism for the recovery of carbon lost from the photosynthetic carbon reduction cycle as a result of oxygenation of RuBP, an important secondary role may be in the prevention of photoinhibitory damage to the photochemical apparatus of the thylakoids. At early stages of leaf development when the capacity for light energy transduction by the thylakoid photochemical apparatus is greater than the capacity for photosynthetic carbon assimilation, photorespiration may be of major importance in dissipating photochemically generated reducing power and thereby preventing photoinhibition of the photochemical apparatus occurring. In order for photorespiration to serve this purpose in the expected fashion as outlined in Fig. 2, it would be necessary for the photosynthetic carbon reduction cycle to be operating and that significant oxygenase and carboxylase activities of RuBPc-o be present.

In the context of leaf development the role of photorespiration in nitrogen metabolism should not be ignored. Nitrogen released in the form of ammonia by the decarboxylation of glycine to serine, an important step in the photorespiratory carbon oxidation pathway which occurs in the mitochondria (Fig. 2), can be reassimilated by the glutamate synthase cycle (Wallsgrove _et al._, 1983). This flux of nitrogen through the photorespiratory and glutamate synthase cycles constitutes a major part of the mature leaf's nitrogen metabolism since it is thought to be greater than ten times the rate of primary nitrogen assimilation in the leaf (Wallsgrove et al., 1983). The glutamate synthase cycle, which involves glutamine synthetase and glutamate synthase, operates in the chloroplast and is light-dependent; photosynthetic electron transport provides ATP for glutamine synthetase and reducing power for glutamate synthase (Wallsgrove et al., 1983). Low levels of these two enzymes have been found in etiolated leaves, but they rapidly increase during greening. Two forms of glutamate synthase exist; a NAD(P)H- and a ferredoxin-dependent enzyme. Interestingly the NAD(P)H-dependent activity can be 20-50% of the total activity in etiolated leaf tissue but constitutes less than 3% in mature leaves (Wallsgrove et al., 1983). This change in the nature of the reducing species required for maximal glutamate synthase activity in the leaf during development may be indicative of the potential supply of the two reducing species available in the plastids during leaf ontogeny. Also, since nitrate and nitrite reduction in leaves is stimulated in the light (Wallsgrove et al., 1983), presumably as a result of increased supply of ATP and reducing power, these activities may provide a sink for excessive reducing power generated by the photochemical apparatus, but not consumed by carbon metabolism, at early stages of leaf growth. Quite clearly there will be a large demand for nitrogen assimilation and metabolism during leaf development and it can be speculated that nitrogen metabolism, possibly in conjunction with photorespiratory activity, may provide the major sink for photosynthetically generated reducing power and ATP. This is an attractive hypothesis since it would explain the high photorespiration rates at early stages of leaf growth and also provide an explanation of how severe photoinhibition of the developing photochemical apparatus may be minimised. As yet there is no substantial experimental evidence to support this hypothesis, however it is to be hoped that this area will become of major interest to physiologists in the future.

CONCLUDING REMARKS

It is evident from the preceding sections of this chapter that it is currently impossible to present a precise summary of the changes in energy metabolism that are occurring during leaf growth. Even if comprehensive information of the changes in the various energy transducing processes was available for leaf growth of a given species, it is likely that this could not be simply extrapolated to predict the situation in leaves of other species. It seems more relevant here to consider some of the major problems that will have to be addressed before a sound understanding of the interactions between the major energy transducing processes in the developing leaf can emerge.

The practical difficulties in isolating metabolically active mitochondria from leaf tissue have seriously restricted studies on respiratory metabolism in the developing leaf. The large changes occurring in respiratory activity at early stages of leaf growth are presumably the result of mitochondrial development and rapidly changing metabolic demands of the growing leaf. The factors controlling the development and activities of the respiratory apparatus are unclear and warrant detailed investigation. Similarly, extensive biochemical studies on the development of the photosynthetic carbon reduction and photorespiratory carbon oxidation pathways are required before an understanding of the changes in energy metabolism that occur during leaf growth can be achieved. It is possible that the metabolic pathways associated with photosynthetic carbon assimilation, photorespiration and respiration may change markedly during leaf development and further confound the situation. Besides clarification of the metabolic pathways involved in energy transduction, it is also essential to obtain information on the rates of metabolite turnover, the size of metabolite pools, and enzyme concentrations, specific activities and characteristics, if the interactions occurring between the metabolic processes are to be resolved. It must also be realised that a major control on metabolism in eukaryotic cells is often affected by compartmentation of reaction sequences. Consequently the rate of flux of metabolites across the membranes separating adjacent compartments may be an important limitation to metabolic activity. Preliminary studies on the phosphate, carboxylate and adenine nucleotide translocators in the mitochondrial and plastid envelopes have suggested that large changes in their activities occur during leaf development (Wellburn, 1982). It may well be, because of the extremely complex and variable nature of energy metabolism during leaf growth, that the application of computer modelling will be required to enable even a basic understanding of the problem to be achieved.

A particularly intriguing feature of the development of the photosynthetic apparatus that warrants further consideration is the asynchronous development of the light-energy transducing photochemical apparatus of the thylakoids and the capacity for carbon assimilation. Not surprisingly in the majority of species studied the development of carbon dioxide photoassimilation appears to be correlated to the synthesis of RuBPc-o. It would be energetically inefficient to synthesise a full complement of RuBPc-o prior to the assembly of the other component enzymes of the carbon reduction cycle, since RuBPc-o constitutes a major proportion of the soluble protein in the mature leaf. The development of a significant rate of carbon assimilation occurs after the formation of a photochemically competent thylakoid population and raises the question of whether the leaf specifically requires a light-dependent supply of reducing power, and possibly ATP, at the early stages of development. Alternatively, the appearance of light-energy transduction prior to carbon assimilation may simply be an inevitable consequence of the leaf having to synthesise some components of the photosynthetic apparatus before others because the biosynthetic machinery required for production of all the components cannot be fully assembled and/or driven simultaneously due to physical and/or energy supply constraints. It is not difficult to envisage nitrogen metabolism within the developing chloroplast acting as a major sink for reducing power and ATP produced by the thylakoids (see Photorespiration section). If this is the case, then the question must be asked as to why reducing power and ATP produced from respiratory activity, which is high at early stages of leaf growth, cannot be translocated into the chloroplast to satisfy such a demand. If the light energy absorbed by the developing thylakoids is not transduced into chemical energy and consumed in metabolism, then a potentially damaging situation arises. An excess of trapped light energy will reduce the photochemical capacity of PS2 and eventually lead to a loss of chlorophyll. Such photodamage can occur under relatively low light levels if there is no significant sink for PS2 to transfer electrons to. Preliminary in vivo studies on developing leaves have not suggested that a large amount of photodamage does occur; this is as might be expeted since it would be wasteful to sythesise the photochemical apparatus only to have it immediately susceptible to severe damage by any light it receives. A more tenable hypothesis would be that the developing leaf is capable of rapidly dissipating any excess reducing power generated by the photochemical apparatus. This may be achieved by the thylakoids performing a State transition or alternatively by the removal of reducing power either via the photorespiratory carbon oxidation pathway or directly to oxygen. Although photorespiration can act as a sink for NADPH and ATP in mature leaves (see Fig. 2; Osmond, 1981), it seems unlikely that a similar situation can exist in the early phases of leaf growth since the absence of significant amounts of RuBPc-o would preclude this. It may be possible for a modified photorespiratory pathway, which would not invoke oxygenase activity of RuBPc-o, to operate in developing leaves and consume reducing power.

It is to be hoped that given the considerable amount of resources and research effort currently being invested in the study of photosynthesis and the associated metabolic processes, new knowledge and techniques will emerge which may help to resolve some of the problems and questions relating to energy transduction during leaf growth which are raised in this chapter. The expanding leaf with its sequential development of respiratory, photorespiratory and photosynthetic activities may prove not only to offer an interesting developmental problem, but may also prove to be an extremely useful tool with which to examine the complex regulatory interactions that can occur between the various energy transducing systems.

ACKNOWLEDGEMENTS

I am grateful to S.P. Long for many helpful discussions on topics relating to this chapter.

REFERENCES

Baker, N.R. (1984). Development of chloroplast photochemical functions. In Chloroplast Biogenesis, Vol. 5, Topics in Photosynthesis, eds. N.R. Baker & J. Barber, pp. 207-52. Amsterdam: Elsevier.

Baker, N.R. & Hardwick, K. (1973). Biochemical and physiological aspects of leaf development in cocoa (*Theobroma cacao*). I. Development of chlorophyll and photosynthetic activity. New Phytol., 72, 1315-24.

Barber, J. (1982). Influence of surface charges on thylakoid structure and function. Ann. Rev. Plant Physiol., 33, 261-95.

Barber, J. (1983). Photosynthetic electron transport in relation to thylakoid membrane composition and organisation. Plant, Cell and Environ., 6, 311-22.

Callow. J.A. (1974). Ribosomal RNA, fraction I protein synthesis and ribulose diphosphate carboxylase activity in developing and senescing leaves of cucumber. New Phytol., 73, 13-20.

Hampp, R. & Wellburn, A.R. (1979). Control of mitochondrial activities by phytochrome during greening. Planta, 147, 229-35.

Haworth, P., Kyle, D.J., Horton, P. & Arntzen, C.J. (1982). Chloroplast membrane protein phosphorylation. Photochem. Photobiol., 36, 743-8.

Hinkle, P.C. & McCarty, R.E. (1978). How cells make ATP. Sci. Amer., 238, 104-23.

Kelly, G.J. (1983). The complexities of respiration in photosynthetic cells. Trends Biochem. Sci., 8, 38.

Kennedy, R.A. & Johnson, D. (1981). Changes in

photosynthetic characteristics during leaf development in apple. Photosynth. Res., 2, 213-23.

Leech, R.M. & Baker, N.R. (1983). The development of photosynthetic capacity in leaves. In The Growth and Functioning of Leaves, eds. J.E. Dale & F.L. Milthorpe, pp. 271-307. Cambridge: Cambridge University Press.

Lorimer, G.H. (1981). The carboxylation and oxygenation of ribulose 1,5-bisphosphate: The primary events in photosynthesis and photorespiration. Ann. Rev. Plant Physiol., 32, 349-83.

Miranda, V., Baker, N.R. & Long, S.P. (1981). Limitations of photosynthesis in different regions of the Zea mays leaf. New Phytol., 89, 179-90.

Osmond, C.B. (1981). Photorespiration and photoinhibition. Some implications for the energetics of photosynthesis. Biochim. Biophys. Acta, 639, 77-98.

Peisker, M., Ticha, I. & Catsky, J. (1981). Ontogenetic changes in the internal limitations to bean-leaf photosynthesis. 7. Interpretation of the linear correlation between CO_2 compensation concentration and CO_2 evolution in darkness. Photosynthetica, 15, 161-8.

Thornber, J.P., Markwell, J.P. & Reinmann, S. (1979). Plant chlorophyll-protein complexes: recent advances. Photochem. Photobiol., 29, 1205-16.

Wallsgrove, R.M., Keys, A.J., Lea, P.J. & Miflin, B.J. (1983). Photosynthesis, photorespiration and nitrogen metabolism. Plant, Cell and Environ., 6, 301-9.

Webber, A.N., Baker, N.R., Platt-Aloia, K.A. & Thomson, W.W. (1984). Appearance of a State 1-State 2 transition during chloroplast development in the wheat leaf: Energetic and structural considerations. Physiol. Plant., 60, 171-9.

Wellburn, A.R. (1982). Bioenergetic and ultrastructural changes associated with chloroplast development. Int. Rev. Cytol., 80, 133-91.

Williams, L.E. & Kennedy, R.A. (1978). Photosynthetic carbon metabolism during leaf ontogeny in Zea mays L.: Enzyme studies. Planta, 142, 269-74.

THE CARBON RELATIONS OF THE DEVELOPING LEAF

J.E. Dale
Department of Botany, University of Edinburgh,
Edinburgh, EH9 3JH, United Kingdom.

INTRODUCTION

From its inception as a primordium, until some time during the phase of lamina expansion, the developing leaf is dependent on import of carbon from other parts of the plant. Even after photosynthetic activity and carbon assimilation have commenced import occurs concurrently and only falls to low and insignificant amounts once lamina expansion is well advanced. This article considers the demand for carbon by the developing leaf and how this is met, together with changes associated with the transition from the importing to the exporting state.

CARBON NUTRITION OF THE PRIMORDIUM

Vascularistion and Import

At inception, leaf primordia are extremely small consisting of a few thousand cells. Because of their size and location, invested as they are by older primordia, young primordia are difficult to study experimentally save by laborious methods of serial sectioning (see Williams, 1975), or by technically-demanding micro-techniques. Using micro-dissection, Sunderland & Brown (1956) were able to determine cell number and size of successive primordia of *Lupinus albus* (Table 1). It is possible to estimate carbon content per primordium and to derive order-of-magnitude values for carbon import, expressed as sucrose equivalents per plastochron. The assumptions made to do this are considerable and since no account is taken of carbon required in respiration the figures should be regarded with caution. Nevertheless the data indicate that just as the

Table 1. The size and carbon and protein-nitrogen contents of successive primordia of Lupinus albus. (After Sunderland and Brown 1956, Sunderland et al., 1957).

Primordium Number	Primordium Volume (μm X10)	Cell Number X10	Cell Volume (μm^3)	Carbon per cell* (pg)	Protein-N per cell (pg)
1	0.4	1.6	230	16	10
2	0.8	2.4	320	23	11
3	1.4	4.3	340	26	10
4	2.5	7.4	340	26	9
5	4.9	12.0	410	32	9
6	10.0	16.3	610	48	14
7	20.9	26.4	790	61	19

* Based on assumptions (1) that density of primordium is 1.0, (2) that water content is 80 per cent, (3) that carbon content is 44 per cent of (dry weight - weight of protein-nitrogen).

primordium enlarges exponentially so too carbon import increases at an exponential or greater than exponential rate (Fig. 1).

Initially, the absolute demand for carbon by the primordium is small but for Lupinus even the youngest may require about 100 ng sucrose equivalent during the first plastochron. Since vascular differentiation is only just beginning at this stage it seems inevitable that much of the carbon imported will move into and through the primordium by diffusion. Larson (1983) has proposed that procambial strands develop acropetally into the growing stem apex in advance of the leaf primordia they are destined to serve. Whether or not the relationship between procambial development and the siting of primordia is causal, as Larson suggests, extension of the procambium into the primordium certainly occurs very early and this has great significance for import. Acropetal differentiation of phloem proceeds

Fig. 1. The import of protein and carbon (estimated as sucrose) by leaf primordia of Lupinus albus expressed per plastochron. (Protein-N values are those of Sunderland et al., 1957, C values are estimated from data in Table 1.)

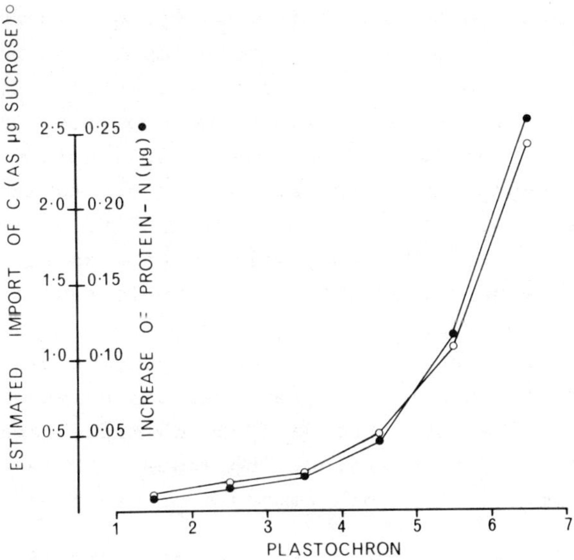

rapidly, in advance of xylem (e.g. Lersten, 1965; Esau, 1965; Coleman & Greyson, 1976), and by the time that the primordium is a few hundred µm in length differentiated phloem has reached the tip and is involved in carbon translocation into and within the primordium. Only with functional phloem can the substantial amounts of carbon required in the later plastochrons be met.

Primordial Metabolism

As the primordium increases in size so its respiration rate increases also, whether expressed on an organ basis or per unit insoluble nitrogen (Sunderland et al., 1957). Some information is now available on the pathway of oxidation of carbohydrate in tissues of the shoot apical meristem and leaf primordia (Croxdale, 1983; Croxdale

& Outlaw, 1983). Very sensitive microassays (Lowry & Passonneau, 1972; Outlaw, 1980) for the enzymes phosphofructokinase, a marker for glycolysis, and glucose-6-phosphate dehydrogenase, a marker for the pentose phosphate pathway (PPP) have shown that glycolysis is the main path for carbohydrate breakdown in very young primordia, but that PPP becomes proportionately more important for primordia more than 2 plastochrons old (Table 2). Since PPP is important for the generation of metabolites such as NADPH, pentoses (required for synthesis of nucleic acids) and erythrose-4-phosphate (required for synthesis of aromatic amino acids), the gradual switch to this mode of carbohydrate breakdown may well be an essential adjunct to the increasing differentiation of cells and tissues in the growing primordium.

The heterogeneity of cell types in older primordia, reflected in their different metabolic rates (Table 2), makes the few published data on chemical composition of primordial tissues difficult to assess. Data of Williams & Rijven (1965, 1970) for primordia of wheat and subterranean

Table 2. Activities of phosphofructokinase (PFK) and glucose-6-phosphate dehydrogenase (G-6-P), expressed as nmol substrate consumed per kg dry matter per h, in primordia of Dianthus chinensis. (After Croxdale, 1983).

Enzyme	G-6-P	PFK
Tissue		
Apex	250	1056
Leaf primordium 1)		
)	550	1000
Leaf primordium 2)		
LP6		
Epidermis	350	140
Mesophyll	350	160
Xylem	350	350
Phloem	725	320
LP9 and older	200	80

clover represent average values for a range of cell types (Table 3) but do nevertheless clearly indicate that cell wall synthesis is a major feature of growing cells. For the fourth leaf of wheat, the average values of cell wall per cell increase in total by about 90-fold prior to the commencement of leaf emergence, and a further 10-fold subsequently.

Table 3. Changes in protein-nitrogen, RNA phosphorus and cell wall, all as pg per cell, for primordia of wheat (Triticum aestivum) and subterranean clover (Trifolium subterraneum) in the 7 days prior to emergence. (Data of Williams and Rijven 1965, 1970).

	Wheat			Subterranean Clover		
Days before emergence	Protein-N	RNA-P	Cell wall	Protein-N	RNA-P	Cell wall
7	56	6.6	46	13	1.0	8
6	73	7.8	106	12	1.2	32
5	80	7.9	90	13	1.4	62
4	77	9.7	109	13	1.3	94
3	70	9.2	183	12	1.5	91
2	80	10.7	300	16	1.7	157
1	110	12.2	537	29	2.5	264

This massive synthesis of carbohydrate polymers in the primordium suggests that much of the import is likely to be of carbohydrate such as sucrose, originating from photosynthetic activity of older leaves. Little is known of the origin of the nitrogen in the primordium and whether nitrate assimilation can occur in primordial tissues has not been investigated. In Lupinus there is an exponential increase in protein-nitrogen per primordium when plotted on a plastochron scale (Sunderland et al., 1957; see also Fig. 1 which plots the increase in protein-nitrogen per plastochron). It seems likely that most nitrogen will be imported by the primordium in the amino or amide form, either directly from older leaves along with assimilate, or indirectly from the roots, but our ignorance on these points is almost total.

THE EXPANDING LEAF

General Features

For most dicotyledon species the phase of lamina unfolding is marked by major increases in both cell number and cell size. Up to 90% of all cells in the leaf may be produced by divisions over this period (Fig. 2; Sunderland, 1960; Milthorpe & Newton, 1963; Dale, 1976) and increases in mean cell volume of 15-22 fold are common (Sunderland, 1960; Dale & Murray, 1968; Verbelen & de Greef, 1979). Invariably there is a period when cell division has ceased but lamina expansion continues as a result of cell enlargement and intercellular space formation. This phase is often short, but it is sufficiently long in primary leaves of *Phaseolus vulgaris* (Dale & Murray, 1968; Verbelen & de Greef, 1979; Fig. 2) for these to be used in studies of leaf cell expansion free of the complication of concomitant cell division (van Volkenburgh & Cleland, 1979, 1980, 1981). In the case of the grass leaf (Fig. 3a) considerable extension growth has to occur before the leaf tip gradually appears above the sheaths of subtending leaves, and in those parts of the blade that have emerged cell enlargement and vascular differentiation are already complete (Davidson & Milthorpe, 1966; Patrick, 1972). In grasses cell division is restricted to the narrow meristem at the base of the developing leaf towards the bottom of the enclosing sheath of older leaves (Esau, 1965; Langer, 1979) so that the major processes of cell increase, growth and maturation precede leaf appearance.

It is clear that maturation of the grass leaf occurs earlier at the tip than at the base. This basipetal maturation is a general finding (Maksymowych, 1973; Dale & Milthorpe, 1983), although in some cases, such as in the primary leaves of *Phaseolus vulgaris*, the basipetal gradient of maturity is very short-lived. At the other extreme is the

Fig. 2. and 3. The time course in cell number, CN, leaf area, A, and net photosynthesis, P (a), import (b), and export (c) of carbon. All values are expressed as a percentage of the maximum. The arrows indicate the periods of carbon import and export. Data in Fig. 2 (left) are for primary leaves of Phaseolus vulgaris; data in Fig. 3 (right) are for the second leaf of Hordeum vulgare.

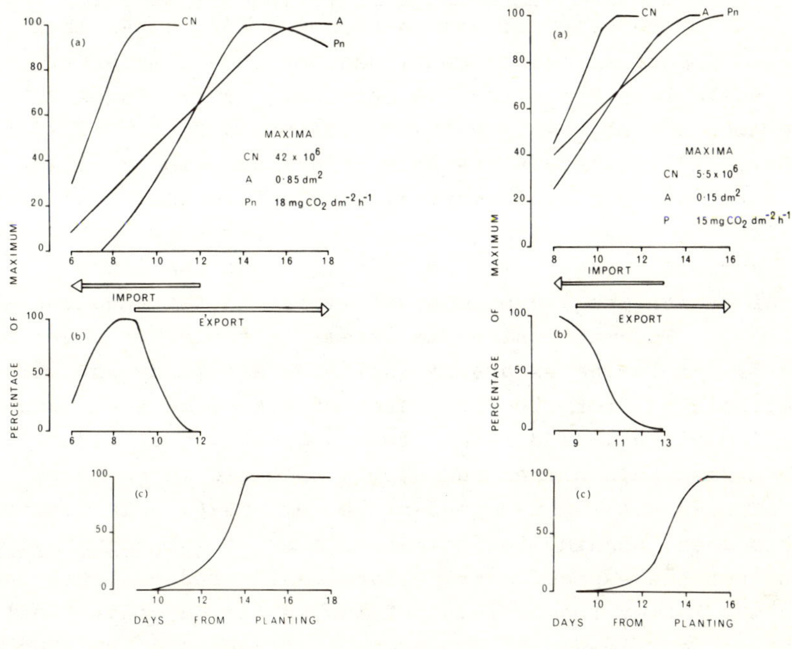

development of the leaf-like cotyledon of unifoliate species of Streptocarpus (e.g. S. grandis, S. wendlandii) where expansion occurs from a basal meristematic region and continues over several months to give a "blade" of up to 500 cm in area (see Jong & Burtt, 1975 for an account of leaf development in this genus).

In terms of the carbon economy of the leaf development of the vascular tissues is of major importance since this must govern rate and pattern of import and export. The general rules are simple to summarise: there is acropetal development of both phloem and xylem in the midrib and high order veins which is largely completed in the primordial phase; during unfolding of the blade, structural and functional maturation of the minor veins proceeds from the leaf tip downwards.

Import: The Sources of Carbon for Growth

Wherever it has been looked for it has been found that import of carbon continues for some time after export of assimilate has begun (Hopkinson, 1964; Thrower, 1967; Turgeon & Webb, 1973, 1975; Fellows & Geiger, 1974; Ho & Shaw, 1977; Anderson & Dale, 1983; see also Fig. 2 and 3). Import of carbon is maximal long before the leaf reaches half its final size, falling to very low rates before expansion ceases. The implication that the final stages of leaf growth are independent of import is returned to later.

The extensive meristematic activity that occurs early in lamina expansion indicates a demand for imported metabolites for the processes of synthesis that accompany cell division, and it has been suggested that older leaves may meet this demand and supply specific metabolites whose synthesis is light-dependent (Dale, 1976). In support of this idea, shading the primary leaves of Phaseolus vulgaris but not the expanding trifoliate leaf, reduces cell number in the latter (Dale, 1976). Likewise, in Cucumis final cell number of leaves is dependent on irradiance (Milthorpe & Newton, 1963); this is explicable if successive leaves produce more of the materials required to prime cell division in the younger leaves they supply in high than in low light. An implication is that expanding leaves are not able to provide enough of the required substrates from local photosynthesis.

Experiments where leaves have been fed with $^{14}CO_2$ have shown source leaves to supply, preferentially, younger leaves in direct vascular connection with them (Joy, 1964; Larson & Dickson, 1973; Turgeon, 1980). The specificity of the source:sink relation may be very strong. For example,

Barlow (1979) showed that feeding a particular source leaf of apple led to accumulation of isotope in only a small area of an expanding sink leaf. In Phaseolus vulgaris when $^{14}CO_2$ is supplied to one primary leaf there is accumulation of label mainly in one lateral of the first trifoliate leaflets (Fig. 4), despite the fact that there are demonstrable vascular paths between each primary leaf and all the leaflets (Borchers-Zampini et al., 1980). Removal of the other primary leaf results in a significant increase in label going to the leaflets normally supplied with only small quantities. In short-term experiments over 6 h, shading the other primary has no effect on carbon partition in the trifoliate, presumably because the shaded leaf can continue to supply the trifoliate from previously-formed assimilate.

Fig. 4. The effect of treatment of a primary leaf of Phaseolus vulgaris on the distribution of labelled carbon fixed in the older leaf and translocated to the developing first trifoliate leaf. (a) Control; $^{14}CO_2$ *fed to arrowed leaf with most carbon translocated to the nearest trifoliate leaflet. (b) One of the primary leaves shaded with aluminium foil from the time the other leaf was fed* $^{14}CO_2$ *until harvest; results are similar to those in (a). (c) One primary leaf removed before the other was fed* $^{14}CO_2$*; label was distributed more evenly among the leaflets of the sink leaf.*

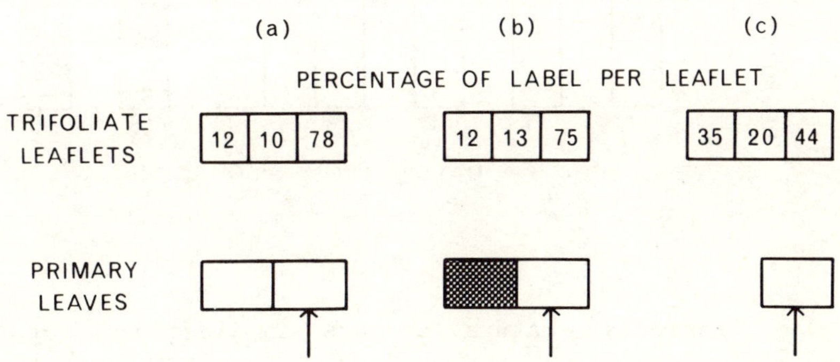

At the other extreme an expanding leaf may be supplied from several older leaves in a complex pattern that changes with time, as seen in data for the expanding third leaf of barley (Fig. 5). Here, prior to emergence of the tip of the expanding leaf most carbon is supplied by the first leaf, but during unfolding the contribution from this leaf gradually falls, to be exceeded by that from the second, and ultimately by that from the third leaf itself. The extensive vascular anastomoses at the nodes of cereal grasses (Patrick, 1972) mean that growing leaves will be supplied from a pool of fixed carbon to which the most actively photosynthesising source leaf will contribute the greatest amount (Anderson & Dale, 1983).

Fig. 5. The relative contribution of carbon from the first leaf (lower open box), second leaf (upper open box) or third leaf (shaded box) to the expanding third leaf of Hordeum vulgare. The upward arrow E marks the onset of leaf unfolding, the downward arrow EC indicates the time when expansion was complete.

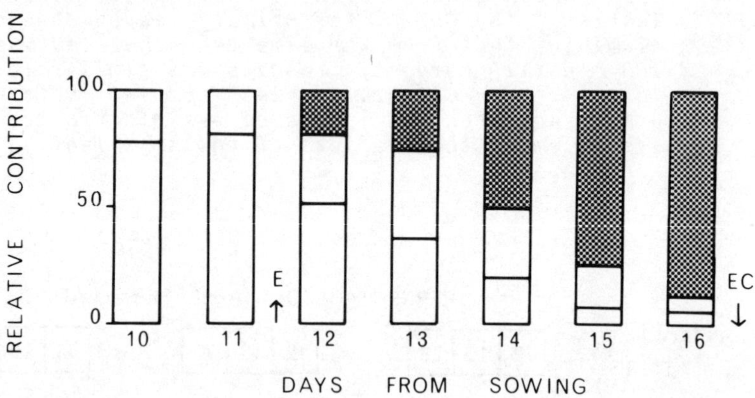

Basipetal maturation has implications for the carbon economy of the leaf which, as it unfolds, begins to photosynthesise. Clearly, carbon requirements can be met by import or by local photosynthesis, or by both. If locally-fixed carbon is used by the leaf in growth this may be done at or close to the site of fixation, or alternatively there may be transport within the developing leaf so that carbon fixed at a more mature distal location is incorporated into expanding cells at the base of the leaf. Although the leaf may import and export carbon concurrently it is unlikely that this will be true also for individual cells in the leaf. For tomato, which has bicollateral bundles, it has been shown that import of carbon occurs mainly through the internal phloem although both internal and external phloem are involved in export (Bonnemain, 1969; Ho & Shaw, 1977). Whether the same sieve elements function for import and later for export in plants without internal phloem is not known, nor is it known whether veinlets of the higher orders, which are the sites of phloem loading (see below), function at any time in carbon import.

The occurrence of such within-lamina redistribution is widespread but not universal. It has been shown for Cucurbita pepo (Turgeon & Webb, 1973, 1975), Pelargonium (Jones & Eagles, 1962) and Lycopersicon (Ho & Shaw, 1977). For the seventh leaf of tomato, the later-maturing basal leaflets can import carbon fixed by the mature terminal ones and there is also preferential import of carbon from older leaves to those leaflets as well. Ho & Shaw (1977) showed that in early leaf expansion more of the imported assimilate is used in protein synthesis than in the manufacture of sugars. However locally-fixed carbon is incorporated into sucrose, and this synthetic activity begins sometime before carbon is exported out of the leaf. In the case of leaves of Populus deltoides carbon fixed by the maturing tip of the blade is found in the developing vascular tissues and the petiole but not in the expanding mesophyll tissues (Larson et al., 1980). In the cereal grasses significant amounts of carbon fixed by the emerged leaf tip can be used in the unemerged expanding basal region (Kemp, 1980a; Anderson & Dale, 1983; Fig. 6). Detailed analysis of sugar content in the expanding region of leaves of field-grown wheat suggest that hexose concentration there is unlikely to limit leaf extension rate (Kemp, 1980b; Kemp & Blacklow, 1981), and that temperature is more likely to be a major determinant of the rate of leaf extension, although not necessarily of final size (Dale, 1982).

Fig. 6. The fate of carbon fixed by the third leaf of Hordeum vulgare at three stages prior to full expansion.

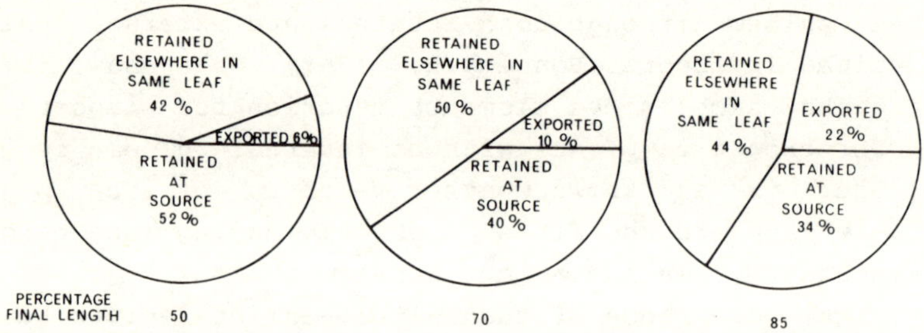

Phloem Loading

A recurring theme of this article is that the growing leaf cannot be regarded as a structurally or physiologically uniform entity with respect to either time or space; at any moment different parts of the expanding leaf will be at different developmental states. Nowhere is this more important than when considering translocation and the export of assimilate, and the point is well made by the fact that distal portions of a developing leaf may export carbon to expanding basal regions.

There are a number of obvious prerequisites before export can occur; there must be anatomical maturation, the phloem-loading mechanism must be functional, and above all, local production of assimilated carbon must exceed local demand so that there is material for export. The broad outlines of the progress of vascularisation in the developing leaf have already been discussed. The acropetal development of the major veins can be considered as a necessity for the nutrition of the leaf in its early development as an importing organ. The basipetal maturation of the higher order veins is closely linked to export.

The sites of carbon fixation in photosynthesis are more-or-less distant from the sites of entry of assimilate into the transport pathway, the sieve elements. Furthermore although sucrose may be present at high concentrations in the photosynthesising cells of the mesophyll, its concentration in the sieve elements is even higher and can be up to 150 kg m^{-3} (15%) in water (Fisher, 1978). Loading into the phloem therefore occurs against a concentration gradient, and cannot be due to diffusion alone. A necessary requirement is that leakage of sucrose out of the phloem is controlled.

Probably the most systematic study of the onset and mechanism of phloem loading is that made of the developing leaves of sugar beet (Geiger & Cataldo, 1969; Geiger et al., 1973; Fellows & Geiger, 1974). Here there is experimental evidence that the loading of assimilate into the phloem is restricted to the sieve element:companion cell complexes (SE:CC) of minor veins of the fifth order. The loading process involves the extrusion of protons from the symplast of the SE:CC and cotransport of sucrose and protons from the surrounding apoplast into the symplast. Evidence for such a proton-sucrose cotransport mechanism comes from several species in addition to sugar beet (see review by Giaquinta, 1983).

The site of entry of assimilate to the apoplast is not known although evidence for it occurring close to the loading minor vein has been reported (Doman & Geiger, 1979). Since no mesophyll cell is very far from a loading site, with 20-30 mesophyll cells supplying a fifth order vein in mature sugar beet leaves, the distance involved for apoplastic transport is unlikely to exceed 100 μm (Geiger, 1975), and distances for inter-cellular symplastic transport will also be correspondingly short.

The Export Phase

For export to occur cells of the mesophyll, as well as being photosynthetically competent, must no longer be net sinks for carbon and it follows that they will have completed much if not all of their growth and the synthesis of the major structural component cell wall, protein, membranes, etc. It is also a requirement that there must be the production of molecules for export, the so-called transfer carbohydrates (Bieleski, 1976), of which sucrose is the most usual. It is not therefore surprising that sucrose sythesis has been shown to precede significant export in a number of species. But high concentrations of sucrose in a leaf do not necessarily mean that export will occur. The transfer carbohydrate must be available for loading into the phloem and must enter the apoplast, reach the site of an active, competent loading system, traversing at least one membrane for this purpose.

Judged by published values of the reflection coefficient, admittedly for cells of Nitella (Steudle & Zimmermann, 1974), the permeability of plant cell membranes to sucrose is quite low. The accumulation of sucrose in fully expanded mesophyll cells is likely to increase cell turgor and if membranes are sensitive to pressure (Zimmermann, 1977) this may lead to a change in permeability characteristics. Alternatively, proton extrusion by the loading mechanism in the phloem could lower the pH of the apoplast and bring about a consequent change in the permeability of adjacent membranes. A third possibility is that an active sucrose extrusion pump develops at the mesophyll cell membrane, "leaking" sucrose into the apoplast. This seems the least likely mechanism as it is energetically less economical than the alternatives. Demonstration that sucrose reaches the apoplast before the loading mechanism is functional would be good but not essential evidence in favour of the turgor hypothesis.

Once the leaf is fully expanded storage carbohydrates such as starch, or fructans in the case of cereal grasses, may begin to be formed. The partition of carbon between storage and export pools is closely controlled. This could be achieved if, when supply of transfer carbohydrate to the phloem exceeds demand elsewhere in the plant, there is a reduction in activity of the loading mechanism with a resultant rise in apoplastic sucrose concentration and reduced efflux from the mesophyll; in turn, the rise in sucrose concentration in the mesophyll leads to the diversion of fixed carbon into the formation of storage carbohydrate, perhaps through a turgor-based osmoregulatory mechanism. Unfortunately the inter-relationships within and between cells in the short-term control of metabolic activity remain under-researched.

CONCLUDING REMARKS

One of the central problems in leaf development is logistical. From primordial initiation until well into the unfolding phase the leaf is dependent on supply of metabolites from other parts of the plant. Rapid development and ramification of the vascular network is essential for this supply, even though in absolute terms import is initially small. Export too is dependent on the maturation and full differentiation of a phloem-loading mechanism, development of structure and biochemical function both being essential. Although the significant developmental changes that occur during leaf expansion with respect to carbon relations can be identified, there remain major uncertainties about the nature of the controlling mechanisms. Development of micro methods for analysing the physical properties and chemical composition of leaf cells will contribute increasingly to our understanding of these mechanisms.

REFERENCES

Anderson, L.S. & Dale, J.E. (1983). The sources of carbon for developing leaves. J. Exp. Bot., 34, 405-14.

Barlow, H.W.B. (1979). Sectorial patterns in leaves on fruit trees produced by radioactive assimilates and solutions. Ann. Bot., 43, 595-602.

Bieleski, R.L. (1976). Transport of sorbitol in pear leaf

slices. In Transport and Transfer Processes in Plants, eds. I.F. Wardlaw and J.B. Passioura, pp. 185-90. New York: Academic Press.

Bonnemain, J.L. (1969). Transport du ^{14}C Assimilé a partir des feuilles de tomato en voie de croissances et vers celles-ci. Comp. Rend. Hebd. Seance Acad. Sci., Paris, 269D, 1660-3.

Borchers-Zampini, C., Glamm, A.B., Hoddinott, J. & Swanson, C.A. (1980). Alterations in source-sink patterns by modifications of source strength. Plant Physiol., 65, 1116-20.

Coleman, W.K. & Greyson, R.I. (1976). The growth and development of the leaf in tomato (Lycopersicon esculentum). II. Leaf ontogeny. Can. J. Bot., 54, 2704-7.

Croxdale, J.G. (1983). Quantitative measurements of phosphofructokinase in the shoot apical meristem, leaf primordia and leaf tissues of Dianthus chinensis L. Plant Physiol., 73, 66-70.

Croxdale, J.G. & Outlaw, W.H. (1983). Glucose-6-phosphate-dehydrogenase activity in the shoot apical meristem, leaf primordia and leaf tissues of Dianthus chinensis L. Plant, 157, 289-97.

Dale, J.E. (1976). Cell division in leaves. In Cell Division in Higher Plants, ed. M.M. Yeoman, pp. 315-45. London: Academic Press.

Dale, J.E. (1982). Some effects of temperature and irradiance on the first four leaves of wheat, Triticum aestivum. Ann. Bot., 50, 851-8.

Dale, J.E. & Milthorpe, F.L. (1983). General features of the production and growth of leaves. In The Growth and Functioning of Leaves, eds. J.E. Dale & F.L. Milthorpe, pp. 151-78. Cambridge University Press.

Dale, J.E. & Murray, D. (1968). Photomorphogenesis, photosynthesis and early growth of primary leaves of Phaseolus vulgaris. Ann. Bot., 32, 767-80.

Davidson, J.L. & Milthorpe, F.L. (1966). Leaf growth in Dactylis glomerata following defoliation. Ann. Bot., 30, 173-84.

Doman, D.C. & Geiger, D.R. (1979). Effect of exogenously supplied foliar potassium on phloem loading in Beta vulgaris L. Plant Physiol., 64, 528-33.

Esau, K. (1965). Plant Anatomy, 2nd edition. New York: John Wiley.

Fellows, R.J. & Geiger, D.R. (1974). Structural and physiological changes in sugar beet leaves during sink to source conversion. Plant Physiol., 54, 877-85.

Fisher, D.B. (1978). The estimation of sugar concentration in individual sieve-tube elements by negative staining. Planta, 139, 19-24.

Geiger, D.R. & Cataldo, D.A. (1969). Leaf structure and

translocation in sugar beet. Plant Physiol., 44, 45-54.

Geiger, D.R., Sovonick, S.A., Shock, T.L. & Fellows, R.J. (1974). Role of free space in translocation in sugar beet. Plant Physiol., 54, 892-898.

Giaquinta, R.T. (1983). Phloem loading of sucrose. Ann. Rev. Plant Physiol., 34, 347-87.

Ho, L.C. & Shaw, A.F. (1977). Carbon economy and translocation of ^{14}C in leaflets of the seventh leaf of tomato during leaf expansion. Ann. Bot., 41, 833-48.

Hopkinson, J.M. (1964). Studies on the expansion of the leaf surface. IV. The carbon and phosphorus economy of a leaf. J. Exp. Bot., 15, 125-37.

Jones, H. & Eagles, J.E. (1962). Translocation of ^{14}carbon between and within leaves. Ann. Bot., 26, 505-10.

Jong, K. & Burtt, B.L. (1975). The evolution of morphological novelty exemplified in the growth patterns of some Gesneriaceae. New Phytol., 75, 297-311.

Joy, K.W. (1964). Translocation in sugar beet. 1. Assimilation of $^{14}CO_2$ and distribution of material from the leaves. J. Exp. Bot., 15, 485-94.

Kemp, D.R. (1980a). The location and size of the extension zone of emerging wheat leaves. New Phytol., 84, 729-38.

Kemp, D.R. (1980b). The growth rate of successive leaves of wheat plants in relation to sugar and protein concentrations in the extension zone. J. Exp. Bot., 31, 1399-1411.

Kemp, D.R. (1981). The growth rate of wheat leaves in relation to the extension zone sugar concentration manipulated by shading. J. Exp. Bot., 32, 141-50.

Kemp, D.R. & Blacklow, W.M. (1980). Diurnal extension rates of wheat leaves in relation to temperatures and carbohydrate concentrations of the extension zone. J. Exp. Bot., 31, 821-8.

Langer, R.H.M. (1979). How Grasses Grow (Second Edition), pp. 1-66. Studies in Biology no. 34. London: Edward Arnold.

Larson, P.R. (1983). Primary vascularization and the siting of primordia. In The Growth and Functioning of Leaves, eds. J.E. Dale & F.L. Milthorpe, pp. 25-51. Cambridge: Cambridge University Press.

Larson, P.R. & Dickson, R.E. (1973). Distribution of imported ^{14}C in developing leaves of eastern cotton-wood according to phyllotaxy. Planta, 111, 95-112.

Larson, P.R., Isebrands, J.G. & Dickson, R.E. (1980). Sink to source transition of Populus leaves. Ber. Deutsch. Bot. Gesell., 93, 79-90.

Lersten, N. (1965). Histogenesis of leaf venation in

Trifolium wormskioldii (Leguminosae). Am. J. Bot., 52, 767-74.
Lowry, O.H & Passonneau, J.V. (1972). A Flexible System of Enzymatic Analysis. New York: Academic Press.
Maksymowych, R. (1973). Analysis of Leaf Development. Cambridge: Cambridge University Press.
Milthorpe, F.L. & Newton, P. (1963). Studies on the expansion of the leaf surface. III. Influence of radiation on cell division and leaf expansion. J. Exp. Bot., 14, 483-95.
Outlaw, W.H. (1980). A descriptive evaluation of quantitative histochemical methods based on pyridine nucleotides. Ann. Rev. Plant Physiol., 31, 299-311.
Patrick, J.W. (1972). Vascular system of the stem of the wheat plant. II. Development. Aust. J. Bot., 20, 65-78.
Steudle, E. & Zimmermann, V. (1974). Determination of the hydraulic conductivity and of reflection coefficients in Nitella flexilis by means of direct cell-turgor pressure measurements. Biochim. Biophys. Acta, 332, 399-412.
Sunderland, N. (1960). Cell division and expansion in the growth of the leaf. J. Exp. Bot., 11, 68-80.
Sunderland, N. & Brown, R. (1956). Distribution of growth in the apical region of the shoots of Lupinus albus. J. Exp. Bot., 7, 126-45.
Sunderland, N., Heyes, J.K. & Brown, R. (1957). Protein and respiration in the apical region of the shoots of Lupinus albus. J. Exp. Bot., 8, 55-70.
Thrower, S.L. (1967). The pattern of translocation during leaf ageing. Aspects of the Biology of Ageing, ed. H.W. Woolhouse. Symposium of the Society for Experimental Biology, 21 483-506. London: Cambridge University Press.
Turgeon, R. (1980). The import to export transition: experiments on Coleus blumei. Ber. Deutsch. Bot. Gesell., 93, 91-7.
Turgeon, R. & Webb, J.A. (1973). Leaf development and phloem transport in Cucurbita pepo; transition from import to export. Planta, 113, 79-91.
Turgeon, R. & Webb, J.A. (1975). Physiological and structural ontogeny of the source leaf. In Phloem Transport, eds. S. Aronoff et al., New York: Plenum Press.
Van Volkenburgh, E. & Cleland, R.E. (1979). Separation of cell enlargement and division in bean leaves. Planta, 146, 245-7.
Van Volkenburgh, E. & Cleland, R.E. (1980). Proton excretion and cell expansion in bean leaves. Planta, 148, 273-8.
Van Volkenburgh, E. & Cleland, R.E. (1981). Control of

light-induced bean leaf expansion: role of osmotic potential, wall yield stress and hydraulic conductivity. Planta, 153, 572-7.

Verbelen, J.P. & de Greef, J.A. (1979). Leaf development of *Phaseolus vulgaris* L. in light and darkness. Am. J. Bot., 66, 970-6.

Williams, R.F. (1975). The Shoot Apex and Leaf Growth. Cambridge: Cambridge University Press.

Williams, R.F. & Rijven, A.H.G.C. (1965). The physiology of growth in the wheat plant. 2. The dynamics of leaf growth. Aust. J. Biol. Sci., 18, 721-43.

Williams, R.F. & Rijven, A.H.G.C. (1970). The physiology of growth in subterranean clover. 2. The dynamics of leaf growth. Aust. J. Bot., 18, 149-66.

Zimmermann, U. (1977). Cell turgor pressure regulation and turgor pressure-mediated transport processes. Integration of Activity in the Higher Plant, ed. D.H. Jennings. Symposium of the Society for Experimental Biology 31, 117-54. London: Cambridge University Press.

ADAPTIVE SIGNIFICANCE OF LEAF DEVELOPMENT AND STRUCTURAL RESPONSES TO ENVIRONMENT

Hamlyn G. Jones
East Malling Research Station, Maidstone,
Kent, ME19 6BJ, United Kingdom.

INTRODUCTION

Leaf development and final morphology depend on genotype, ontogenetic position and on environment. All features of leaf structure and physiology, including size, shape, cell size, trichrome density, stomatal distribution, photosynthetic capacity and so on are sensitive to environment and contribute to a plant's "fitness". Although a given set of characters may be adaptive in the natural situation, they may have little or no advantage when crop species are introduced into new environments. It is of considerable practical interest to determine how the optimal combination of characters might change, both with changes in environment and, more importantly perhaps, with the elimination of interspecific competition.

In this paper I describe some features of leaf developmental responses to environment, with particular reference to changes in stomatal distribution. Analysis of the advantages of specific distribution patterns will be used to illustrate the techniques available for determining the adaptive advantages of particular characters.

DEVELOPMENTAL EFFECTS ON LEAF MORPHOLOGY

Stomatal Distribution

Research into ontogenetic and environmental effects on stomatal distribution has been reviewed by Ticha (1982). Omitting primordial development (before leaf unfolding), the general effects may be summarised:

a) changes during development of a leaf. Changes in stomatal distribution during leaf development tend to be very consistent over a wide range of species and environmental conditions. Initiation of stomata may continue until the leaf has reached between 10 and 60% of its final size, so that total numbers of stomata and the frequency of stomata per unit area both increase up to that point. After then the stomatal frequency declines as leaf expansion is completed. Development of stomata on adaxial and abaxial leaf surfaces tends to occur in parallel (for some exceptions see below). Since the exposed surface of grass leaves is more or less fully expanded at the time they emerge from the leaf sheath, changes in stomatal density are small compared with those in dicotyledonous species. Stomatal size generally increases during leaf development.

b) Plant age and leaf insertion level. Stomatal distribution on similar leaves does not generally change much with plant age. Beakbane & Mujamder (1975), for example, found similar stomatal frequencies on leaves of second-year shoots of apple trees between 2 and 22 years old. Stomatal size and frequency, as well as many other morphological characters, are, however, very dependent on insertion level. The tendency towards more "xeromorphic" leaves with height (increased frequency of stomata, trichromes and veins; smaller stomata, leaves, cells and airspaces; thicker cell walls and more developed surface wax) is well known (e.g. Maximov, 1929). Although partially genetically determined, these insertion gradients are greater in more xeric conditions (Ticha, 1982).

c) Environmental conditions. Stomatal frequency tends to increase with irradiance during growth. A similar increase is often, but not always, apparent with water stress caused by either dry air or soil. The variable response arises because of two conflicting effects, a decrease in cell size with water stress and a decreased stomatal index (stomata as a proportion of total epidermal cells), which may act to cancel each other out (Quarrie & Jones, 1977). Although water stress does not often affect the ratio of adaxial to abaxial frequency (Grace & Russell,

1977; Quarrie & Jones, 1977), there are several reports of this ratio increasing with irradiance during growth (Gay & Hurd, 1975; Ticha, 1982). High wind speed has also been reported to increase this ratio by more than two-fold in _Festuca_ (Grace & Russell, 1977).

As well as large phenotypic effects on stomatal frequency, size and distribution over the leaf surfaces, there are major genetic effects with both frequency and size differing between species by at least an order of magnitude (Willmer, 1983). Some species are hypostomatous with stomata only on the lower (usually abaxial) leaf surface, others are amphistomatous (similar numbers on each surface), yet others are intermediate between these extremes.

Other Characters

Almost all other leaf characters are equally sensitive to environment. For example, it is well known that other aspects of leaf morphology are sensitive to irradiance with characters such as thickness, internal surface area and pubescence all increasing with irradiance, while cell and leaf size decrease (Esau, 1965; Nobel _et al_., 1975; Smith & Nobel, 1978).

MECHANISMS INVOLVED IN DEVELOPMENTAL EFFECTS ON LEAF MORPHOLOGY

Although it is not the purpose of this paper to review the mechanisms involved in the control of leaf morphology, it is pertinent to consider at a simple level the types of control that exist. In most cases environmental control of leaf development acts through the normal hormonal mechanisms but, in some cases, such as the direct reduction of cell expansion by low turgor pressures during severe water stress, physical processes may be directly involved. Even in the case of water deficits, however, it is likely that many of the leaf developmental changes are mediated by the enhanced concentrations of abscisic acid (ABA) that occur with stress (Jones, 1980; Walton, 1980). The evidence for this comes largely from observations that exogenous applications of ABA can mimic many of the known developmental effects of water stress.

Both water deficits and exogenous ABA application lead to a marked reduction in leaf size with both cell expansion and cell division being affected (Quarrie & Jones, 1977; Walton, 1980; Van Volkenburgh & Davies, 1983). Similarly, increases in trichome frequency and decreases in stomatal index have been reported in wheat in response to both water deficits and ABA treatment (Quarrie & Jones, 1977). There is also evidence that leaf wax production increases with both water stress (Baker, 1974; Bengtson *et al*., 1978) and ABA application (Quarrie & Jones, unpublished data). Application of ABA has also been reported to change the course of leaf development in *Potamogeton* from the submerged to the floating type of leaf, which has cuticle and stomata (Anderson, 1978), and to stimulate the production of spines in cacti (Merida & Arias, 1979). All these responses may be regarded as adaptations favouring drought tolerance, whether by physiological mechanisms acting through water use or leaf temperature relations (Jones, 1983), or by preventing grazing, as may be the case with spines in cacti.

Other leaf developmental changes are also hormonally controlled. For example, the increase in leaf size and the tendency towards entire leaves with decreased lobing as irradiance decreases (Lewis, 1972; Sharma & Sen, 1971) can be mimicked by gibberellin treatment (Gray, 1957; Arnosis *et al*., 1977). In extreme cases such as *Solanum surattense* the leaves may be almost entire if grown in the shade, but small and heavily lobed with prominent spines when grown at high irradiances (Sharma & Sen, 1971). These differences suggest that changes in the balance between endogenous levels of at least gibberellins and ABA are involved.

Several of the developmental responses to increased irradiance and water deficits are similar, *e.g*. smaller leaves and increased pubescence, and may well result from the same mechanism, perhaps involving lowered leaf water potential and increased ABA production, but it is clear that many photomorphogenic responses are more complex (Shropshire & Mohr, 1983). It is difficult to see how localised changes, such as are implied by alterations in the ratio of adaxial to abaxial stomatal frequency, could be explained by general changes in hormone levels.

ADAPTIVE SIGNIFICANCE OF STOMATAL DISTRIBUTION

Three main approaches (correlation, experimentation and modelling) are available for the elucidation of the adaptive advantages of particular developmental or morphological characters. These will be illustrated with reference to attempts to answer the following questions: (i) under what conditions would one expect leaves with stomata on only one surface to have an advantage? (ii) when presence on only one surface is favoured, why is that nearly always the lower surface? (iii) what are the relative advantages of high and low stomatal frequencies? Recent discussion of the adaptive nature of other features of leaf morphology may be found elsewhere (e.g. Lewis, 1972; Parkhurst & Loucks, 1972; Quarrie & Jones, 1977; Grace, 1983; Jones, 1983; Nobel, 1983).

Correlation and Regression

One approach is to correlate for a large number of species the presence of a character (such as hypostomaty) with features of their natural habitats (such as aridity, or light climate), or with morphological or taxonomic features of the plants. Unfortunately, both amphistomatous and hypostomatous species are widely distributed so that it has not been possible to relate stomatal distribution unambiguously to any environmental feature. For example, although data for woody plants and for three Indiana plant families indicated a tendency for the proportion of amphistomatous species to increase in increasingly xeric habitats, this was not true among the *Salicaceae* of Indiana (Parkhurst, 1978). Many other exceptions exist. Mott *et al.* (1982) have suggested that amphistomaty is primarily related to full-sun, high-light habitats, which are, of course, often relatively arid. Korner *et al*. (1979) presented data showing an increased tendency to amphistomaty, together with increased stomatal numbers and conductance, in *Vaccinium myrtilis* with increasing altitude and irradiance.

The correlation approach is best where a quantitative character is being studied. Parkhurst (1978) only used presence of absence of stomata on the adaxial leaf surface, but more information could be obtained by using a character such as the ratio of adaxial to abaxial stomatal frequencies and regressing this on environment.

Unfortunately, there is a strong tendency for different environmental variables to be correlated. For example, typical values of the correlation coefficients relating temperature and radiation, temperature and air humidity deficit, and radiation and humidity deficit are as high as 0.595, 0.713 and 0.816, respectively, during summer in southern England (Jones & Higgs, 1982). This effect, together with the complex interactions between different effects, means that it is difficult, even using multiple regression, to isolate those factors primarily concerned in favouring any particular stomatal distribution.

Instead of a direct relation between hypostomaty and environment, it has been suggested that the primary factor determining stomatal distribution might be morphological. In fact, both Parkhurst (1978) and Mott *et al*. (1982) have presented evidence for amphistomaty increasing with leaf thickness, though again the association is not always good. Certainly the thickest leaves, as in succulents with Crassulacean Acid Metabolism, tend to be amphistomatous, while vertically orientated leaves such as Australian schlerophylls and other isolateral leaves tend to be amphistomatous. Other relationships to life-form are well known, particularly the dominance of complete hypostomaty in temperate deciduous trees, and its rarity in the grasses.

Experimental Methods

Experimental manipulation of the environment can be

used as a tool to study developmental responses. Such studies (e.g. Gay & Hurd, 1975; Ticha, 1982) have indicated a tendency towards hypostomaty as irradiances are lowered, but it does not necessarily follow that this response is adaptive, though one might expect evolution to have favoured the occurrence of adaptive responses.

The use of isogenic lines provides an alternative approach. For example, barley lines differing in stomatal frequencies, but otherwise genetically similar, have been used to determine whether selection for low stomatal frequency favours drought tolerance by decreasing water loss (Jones, 1977). That study did not demonstrate a positive correlation between frequency and transpiration because increasing stomatal pore and leaf size tended to compensate for any effect of reduced frequency. Unfortunately, isogenic lines differing in degree of hypostomaty are not available, so that it is not possible to determine directly those conditions where hypostomaty is advantageous.

Another approach could be to compare the performance of normal amphistomatous leaves and leaves where one side has been sealed, e.g. with vaseline.

Modelling

The most powerful tool for studying adaptive advantages of different leaf characters is provided by modelling. The value of any results, however, depends on selecting the right processes to model, i.e. those that are rate limiting or that form the major selection pressure, as well as on the adequacy of the models themselves.

I will now investigate the conditions that would be expected to favour amphistomaty or hypostomaty, using simple models where the different possible mechanisms are treated separately.

The first two mechanisms are based on the hypothesis that the main adaptive advantage of one leaf type would depend on it conferring a greater assimilation rate for a given water loss (Cowan, 1977; Parkhurst, 1978). As a base for these models I have assumed that the amphistomatous and hypostomatous leaves being compared have identical water vapour resistances, and that the upper and lower surface resistances are equal in the amphistomatous case. Using the networks in Fig. 1, and working with resistances (rather than conductances) for algebraic convenience, it follows from the usual rules, that

$$r_{lw}^{H} + r_{aw} = 0.5\ r_{lw}^{A} + 0.5\ r_{aw} \qquad (1)$$

so that

$$r_{lw}^{A} = 2\ r_{lw}^{H} + r_{aw} \qquad (2)$$

where A and H refer, respectively, to amphistomatous and hypostomatous, the subscripts l and a refer, respectively, to leaf and boundary layer and w refers to water vapour. It is apparent from equation (2) that the leaf resistance, primarily determined by the stomata, is a smaller proportion of the total in the hypostomatous case.

It is worth noting here that since water vapour and heat transfer resistances are defined as identical for each leaf type, there will be no effect of stomatal distribution on leaf energy balance or on evaporation rate, but see equation (4) below. It should, of course, be noted that if one takes the alternative approach of defining the stomatal component of the water vapour resistances as equal for hypostomatous and amphistomatous leaves, then water vapour and heat transfer resistances are no longer identical for each leaf type (*e.g.* Cowan & Troughton, 1971; Jones & Slatyer, 1972; Jarvis & McNaughton, 1985).

Effect of Transfer Coefficients. The different conversions between the transfer coefficients for carbon dioxide and water in the boundary layer and through the stomata (Monteith, 1973; Jones, 1983) can cause the hypostomatous and amphistomatous leaves to have different carbon dioxide resistances, and hence photosynthetic rates. Assuming that $r_{ac}/r_{aw} \simeq 1.42$, as expected for a

Fig. 1. Gas phase resistances in (a) hypostomatous, r^H, and (b) amphistomatous, r^A, leaves. Where upper and lower resistances are equal, k = 1.

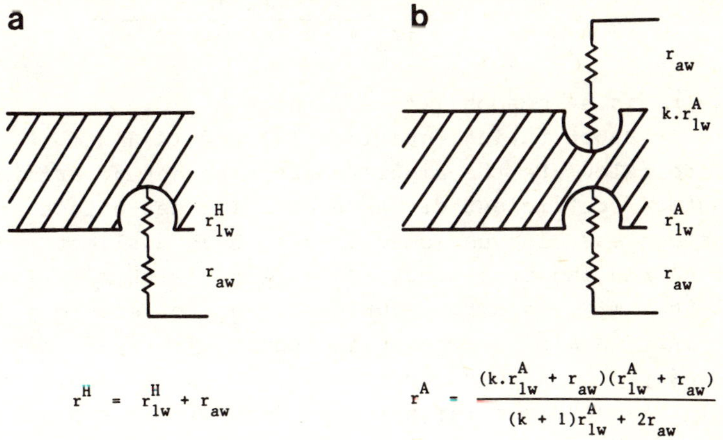

laminar boundary layer, and that $r_{lc}/r_{lw} \simeq 1.64$ (Jones 1983), the total gas-phase carbon dioxide resistances, excepting those in the intercellular spaces, which will be included in the mesophyll resistance for this particular analysis, would be for the hypostomatous leaf:

$$r_c^H = 1.64\, r_{lw}^H + 1.42\, r_{aw} \tag{3}$$

and for the amphistomatous leaf, after substituting for r_{lw}^A and simplifying:

$$r_c^A = 1.64\, r_{lw}^H + 1.53\, r_{aw} \tag{4}$$

One can define a relative amphistomatous advantage, AA, as:

$$AA = (P^A - P^H) / P^H \tag{5}$$

where P^A and P^H, respectively, are the net photosynthetic rates of amphistomatous and hypostomatous leaves. Using the approximation that assimilation is inversely proportional to the total carbon dioxide resistance, though this is only strictly valid where photosynthesis is carbon dioxide limited (Jones, 1983), this gives:

$$AA = \frac{(r_m^H + r_{lc}^H + r_{ac}^H) - (r_m^A + r_{lc}^A + r_{ac}^A)}{(r_m^A + r_{lc}^A + r_{ac}^A)} \qquad (6)$$

where r_m is a "mesophyll" resistance.

Substituting equations (3) and (4) into (6) enables one to calculate the amphistomatous advantage arising from transfer coefficient differences. The net effect is that hypostomatous leaves would tend to have a slight advantage with AA ranging from about -4% (e.g. with r_a = 200 s m^{-1}, r_l = 50 s m^{-1}, and r_m = 200 s m^{-1}) up to zero (e.g. when r_a is very small). The high boundary layer resistances necessary to favour hypostomatous leaves would be most likely to occur for large leaves deep within a plant canopy where windspeeds are lowest. Other effects on AA are summarised in Table 1.

Transfer Through the Mesophyll. Models including gas-phase carbon dioxide transfer within the mesophyll range from simple lumped-resistance models (e.g. Jones & Slatyer, 1972; Parkhurst, 1978) to more complex distributed (Parkhurst, 1977) or continuous (e.g. Jones, 1972; Rand, 1978) models. The simple model shown in Fig. 2 (see Jones & Slatyer, 1972 for discussion and analysis) is adequate to determine the general effect of the intercellular space resistance, r_x, on amphistomatous advantage, using the approach outlined above.

If we eliminate the transfer coefficient effect by setting r_a = 0, the dependence of amphistomatous advantage on mesophyll thickness (m) may be investigated. Estimating r_x as $0.5m/(p.D_c)$ (Parkhurst, 1978), where p is a porosity (assumed = 0.3) and D_c is the molecular diffusivity of carbon dioxide in air (= 14.7mm s^{-1}), r_x would vary between about 6 s m^{-1} for a 50 μm thick leaf to 60 s m^{-1} for a 500 μm leaf. Calculations show that AA would always be positive and increase wih leaf thickness, reaching +10% for a 500 μm leaf with r_i = 200 s m^{-1} and r_{lw}^H = 50 s m^{-1}, but would decrease with increasing r_i and r_l (Table 1).

Table 1. Likely extreme effects of different mechanisms on amphistomatous advantage (AA) using assumptions given in the text.

Mechanism	Value of AA (%)	Size of Effect decreases with	increases with
1. Transfer coefficients	- 4	r_l, r_m	r
2. Mesophyll transfer	+10	r_l, r_m, r_a	leaf thickness
3. Maximum photosynthesis	+20	r_m, r_a	irradiance, H_2O
4. Gradients		r_l, r_m, r_a	
Water vapour	- 2		humidity gradient
Carbon dioxide	- 0.5*		CO_2 gradient
Leaf temperature	- 6		leaf thickness
5. Others			
Protection from dust	- ?		
Air flow through leaf	+ ?		

* upper surface favoured

Extension of the two models given above to leaves intermediate between hypostomatous and equally amphistomatous gives intermediate results.

<u>Effect on Maximum Potential Assimilation Rate</u>. Mott <u>et al</u>. (1982) suggested that the advantage of amphistomaty could be the higher leaf conductance, and hence assimilation rate that would be attainable, but did not attempt to quantify this effect. Evidence collated by Korner <u>et al</u>. (1979) indicates that the minimum leaf resistance <u>per surface</u> is similar for hypostomatous and amphistomatous leaves at about 50 s m , so that amphistomatous leaves have the potential for higher photosynthetic rates. This could be advantageous in conditions of ample water supply. For r_m = 200 s m^{-1}, r_{ac} = 50 s m^{-1} and $r_l^H = r_l^A$ = 50 s m^{-1} this gives a potential amphistomatous advantage of 20%, decreasing with increasing r_m, and increasing with r_a

(Table 1). Because the photosynthetic system would be saturated with carbon dioxide in low light conditions, this effect would be maximal only at high light.

Fig. 2. (a) Model for mesophyll resistance where the photosynthetic tissue is divided into two halves separated by gas-phase diffusion resistances (after Jones & Slatyer, 1972). (b) The same model transformed with the delta-wye transform. Total resistance for k = 1.

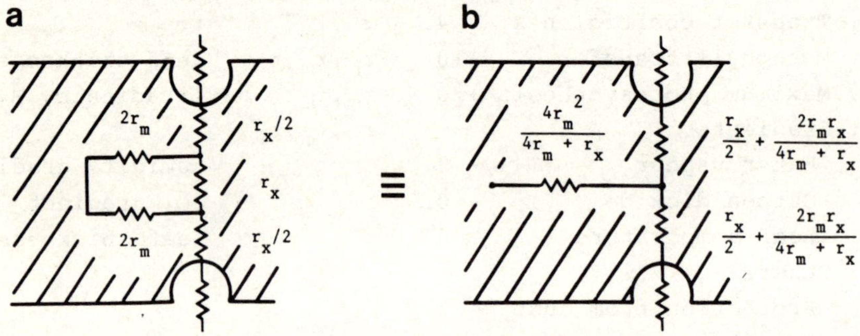

$$r^H = \frac{4r_m^2}{4r_m + r_x} + \frac{2r_m r_x}{4r_m + r_x} + \frac{r_x}{2} + r_{1c}^H + r_{ac} \qquad r^A = \frac{4r_m^2}{4r_m + r_x} + \frac{r_m r_x}{4r_m + r_x} + \frac{r_x}{4} + \frac{r_{1c}^A}{2} + \frac{r_{ac}}{2}$$

Environmental Gradients. So far all the possibilities considered have been non-directional (not distinguishing between hypo- and hyper-stomatous leaves) and have indicated only limited conditions where single sided leaves would have any advantage, and do not appear adequate to explain the widespread occurrence of hypostomaty. Probably, therefore, any advantage of hypostomaty lies not in the fact that stomata are on one side *per se*, but that stomata on the lower surface are favoured. We can investigate the implications of environmental gradients using the hypothesis that the best stomatal distribution gives maximal assimilation per unit water loss. An indication of the effect can be obtained by substituting appropriate values for humidity and carbon dioxide observed above and below the leaves, and for saturation vapour pressures of the two surfaces. This enables one to calculate, using the usual equations (Jones, 1983) but ignoring the feedback imposed by the environment, evaporation and photosynthesis rate for each surface separately.

Humidity tends to increase with depth in the canopy, with the difference from top to bottom of a canopy being frequently of the order of 0.2-0.3 kPa and reaching 1 kPa (equivalent to up to a 90% change) in grassland (Monteith, 1976). Assuming a gradient of only 50 Pa per leaf, this would lead to evaporation from a hypostomatous leaf, E^H , being about 2% lower than E^A at 20°C and 50% relative humidity. Gradients of carbon dioxide concentration are relatively much smaller (Monteith, 1976), with concentrations usually changing by less than 15 $\mu l\ l^{-1}$ through the whole canopy (less than a 5% change). There is a small concentration decrease at the top of the canopy that would tend to favour stomata on the top surface (Table 1), but this effect would normally be much less than 0.5% per leaf, and the gradient reverses deeper in the canopy, favouring hypostomatous leaves deep in the canopy.

Any temperature gradient across leaves would also tend to favour hypostomaty, since the leaf to air vapour pressure gradient would tend to be smaller for the cooler lower surface. Sheriff (1979) has indicated that the top surface may be 1°C warmer than the lower. This would favour hypostomatous leaves as E^H would be up to 6% less than E^A at 50% relative humidity and 20°C. The requisite temperature difference would be most likely to occur with thick leaves (Table 1).

It is worth noting here that many species, particularly among the grasses, have more stomata on the adaxial than on the abaxial surface (e.g. Quarrie & Jones, 1977; Grace & Russell, 1977). Interestingly, however, the leaves of these species are often twisted (Ledent, 1976) such that the abaxial surface is predominantly uppermost in the natural environment (Jones, 1979). In addition, in many of those species leaves roll when stressed, with the adaxial surface innermost.

Other Factors. It is possible that the preponderance of

stomata on the lower surface may be simply an adaption preventing clogging of the pores by raindrops or by dust, or one preventing infection by fungal pathogens (Parkhurst, 1978). In other cases, the occurrence of stomata in localised regions or on a particular surface is correlated with a marked leaf rolling or folding response to drought, as in marram grass, thus protecting the stomata in severe stress. Although the advantage of localised stomatal distribution in such cases is clear, it is difficult to quantify without data on the maximum water vapour conductance of closed stomata compared with that of the folded leaf. Others (Parkhurst, 1978) have suggested that an advantage of amphistomaty may be the possibility for reduction of r_x by bulk air flow through the leaf.

Contrary to initial expectations, there is a clear tendency for stomatal frequency to increase with aridity of environment, both within one plant and between different plants or even species. One possible explanation has already been suggested by the isogenic line studies referred to above and that is that the reductions in cell and leaf size with stress are more important for water conservation or drought tolerance than is a reduction in frequency itself. In fact there is no consistent relation between transpiration and any individual leaf morphological character (Jones, 1985) with, for example, different high transpiring Populus clones having either large stomata or a high stomatal frequency (Siwecki & Kozlowski, 1973). Stomatal aperture is, of course, more important than morphological characters such as stomatal size and frequency in determining transpiration. The fact that stomatal index is reduced, both by water stress and ABA application (Quarrie & Jones, 1977) does suggest that there is a slight advantage in reducing stomatal frequency, other things being equal, but that this advantage is small. The maintenance, or even increase, of stomatal frequency with aridity could even be a positive adaption if, for example, smaller stomata were able to close more effectively than large ones.

OPTIMISATION

Although so far we have treated individual characters separately, the value of any character, such as hypostomaty, depends on the whole system including leaf size, photosynthetic system and environment. A simple understanding of the separate mechanisms involved in plant adaption, as provided by the above models, can be very useful in understanding adaption, but plant breeders and agronomists need to go further. They want to be able to design an ideal plant (ideotype) for any particular situation, or at least to determine the way in which crop adaption could best be improved either by genetic means or by agronomic means, such as by the use of growth regulators.

Development of an ideotype requires that the balance between expression of the various characters is optimised so as to give maximum economic yield. Various approaches to the determination of optimal combinations of plant characters have been reported, particularly in relation to stomatal behaviour (Jones, 1976, 1981; Cowan, 1977, 1983). Although it is easy to determine the stomatal conductance that maximises instantaneous water use efficiency, in the long term the optimum depends partly on the probability of future rainfall and partly on the "strategy" adopted by the plant (Jones, 1981; Cowan, 1983). When one considers morphological characters, the problems are greater because the responses are relatively long term and irreversible, in comparison with stomatal aperture, so that the "cost" of an incorrect response may be greater in terms of lost production or even survival.

Trees tend to be relatively conservative or pessimistic in their responses with a strong emphasis on long-term survival of the individual at the expense of short-term productivity, in contrast with ephemeral species. As pointed out by Mott _et al_. (1982), relatively optimistic plants do tend to have a greater proportion of species with amphistomatous leaves, as would be expected if amphistomaty were, in fact, an adaption for high maximum photosynthesis.

Interestingly, all the simple models described above indicate that one or other of the extremes of amphistomaty or hypostomaty would be expected to be favoured in any particular situation, yet intermediate distributions are common. One possible explanation is that it may be better, in the long run, to be reasonably well adapted for a range of possible future environments than to be ideally adapted for one. It is also likely that the balance between conflicting requirements, such as the advantages of a maximal photosynthetic rate, given by amphistomaty and thick leaves, and the water conservation advantages of hypostomaty may lead to an intermediate distribution.

CONCLUSIONS

This paper has outlined some of the ways in which leaf morphology is dependent on environment, with particular reference to environmental effects on stomatal distribution. The possible adaptive advantages of these different stomatal distributions have been investigated, with the results of applying a range of models being summarised in Table 1. The relative importance of the different effects is not, however, necessarily related to the potential maximum AA. For example, although the maximum photosynthesis effect is potentially the largest, the appropriate conditions (ample water supply and high light) may not be as frequent as conditions for other mechanisms.

REFERENCES

Anderson, L.W.J. (1978). Abscisic acid induces formation of floating leaves in the heterophyllous aquatic angiosperm *Potamogeton nodosus*. Science, 201, 1135-8.

Arnosis, P.A., Caso, O.H. & Cagliatti, D.H. (1977). Efecto de GA y la intensidad de luz sobre la morfogenesis foliar de *Taraxacum officinale*. Phyton, 35, 1-9.

Baker, E.A. (1974). The influence of environment on leaf wax development in *Brassica oleracea* var. *gemmifera*. New Phytol., 73, 955-66.

Beakbane, A.B. & Mujamder, P.K. (1975). A relationship between stomatal density and growth potential in apple rootstocks. J. Hort. Sci., 50, 285-9.

Bengtson, C., Larsson, S. & Liljenberg, C. (1978). Effects of water stress on cuticular transpiration rate and amount and composition of epicuticular wax in seedlings of six oat varieties. Physiol. Plant., 44, 319-24.

Cowan, I.R. (1977). Stomatal behaviour and environment. Advances in Botanical Research, 4, 117-228.

Cowan, I.R. (1983). Regulation of water use in relation to carbon gain in higher plants. In Physiological Ecology II, Encyclopedia of Plant Physiology New Series 12B, eds. O.L. Lange, P.S. Nobel, C.B. Osmond & H. Ziegler, pp. 589-613. Berlin-Heidelberg-New York: Springer-Verlag.

Cowan, I.R. & Troughton, J.M. (1971). The relative role of stomata in transpiration and assimilation. Planta, 97, 325-36.

Esau, K. (1965). Plant Anatomy (2nd ed.). New York: Wiley.

Gay, A.P. & Hurd, R.G. (1975). The influence of light on stomatal density in the tomato. New Phytol., 75, 37-46.

Grace, J. (1983). Plant-Atmosphere Relationships. London: Chapman and Hall.

Grace, J. & Russell, G. (1977). The effect of wind on grasses. III. Influence of continuous drought or wind on anatomy and water relations in *Festuca arundinacea* Schreb. J. Exp. Bot., 28, 268-78.

Gray, R.A. (1957). Alteration of leaf size and shape and other changes by gibberellins in plants. Am. J. Bot., 44, 674-82.

Jarvis, P.G. & McNaughton, K.G. (1985). Stomatal control of transpiration. In Stomatal Function, eds. E. Zeiger, G.D. Farquhar & I.R. Cowan, in press. Stanford: Stanford University Press.

Jones, H.G. (1972). Effects of water stress on photosynthesis. Ph.D. thesis, Australian National University, Canberra.

Jones, H.G. (1976). Crop characteristics and the ratio between assimilation and transpiration. J. Appl. Ecol., 13, 605-22.

Jones, H.G. (1977). Transpiration in barley leaves with differing stomatal frequencies. J. Exp. Bot., 28, 162-8.

Jones, H.G. (1979). Stomatal behaviour and breeding for drought resistance. In Stress Physiology of Crop Plants, eds. H. Mussell & R.C. Staples, pp. 408-28. New York: Wiley.

Jones, H.G. (1980). PGRs and plant water relations. In Aspects and Prospects of Plant Growth Regulators, ed. B. Jeffcoat, pp. 91-100. Letcombe: British Plant Regulator Group.

Jones, H.G. (1981). The use of stochastic modelling to study the influence of stomatal behaviour on yield-climate relationships. In Mathematics and Plant Physiology, eds. D.A. Charles-Edwards & D.A. Rose, pp. 231-44. London: Academic Press.

Jones, H.G. (1983). Plants and Microclimate. Cambridge: Cambridge University Press.

Jones, H.G. (1985). Breeding for stomatal characters. In Stomatal Function, eds. E. Zeiger, G.D. Farquhar & I.R. Cowan, in press. Stanford: Stanford University Press.

Jones, H.G. & Higgs, K.H. (1982). Surface conductance and water balance of developing apple (Malus pumila Mill.) fruits. J. Exp. Bot., 33, 67-77.

Jones, H.G. & Slatyer, R.O. (1972). Effects of intercellular resistances on estimates of the intracellular resistance to CO uptake by plant leaves. Aust. J. Biol. Sci., 25, 443-53.

Korner, Ch., Schel, J.A. & Bauer, H. (1979). Maximum leaf diffusive conductance in vascular plants. Photosynthetica, 13, 45-82.

Ledent, J.-F. (1976). Beam light interception by twisted leaf surfaces. Agric. Meteor., 17, 271-80.

Lewis, M.C. (1972). The physiological significance of variation in leaf structure. Science Progress, 60, 25-51.

Maximov, N.A. (1929). The Plant in Relation to Water. London: Allen and Unwin.

Merida, T. & Arias, I. (1979). Physiological and ecological studies of plant species from the arid and semiarid regions of Venezuela. III. Effects of abscisic acid on spine development in seedlings of Cereus griseus Haw. (Cactaceae). Acta Cien. Venez., 30, 162-6.

Monteith, J.L. (1973). Principles of Environmental Physics. London: Edward Arnold.

Monteith, J.L. (1976). Vegetation and the Atmosphere, Vol. II. Case Studies. London: Academic Press.

Mott, K.A., Gibson, A.C. & O'Leary, J.W. (1982). The adaptive significance of amphistomatic leaves. Plant, Cell & Environ., 5, 455-60.

Nobel, P.S. (1983). Biophysical Plant Physiology and Ecology. San Francisco: Freeman.

Nobel, P.S., Zaragoza, J.L. & Smith, W.K. (1975). Relation between mesophyll surface area, photosynthetic rate, and illumination level during development for leaves of Plectranthus parviflorus Henckel. Plant Physiol., 55, 1067-70.

Parkhurst, D.F. (1977). A three-dimensional model for CO_2 uptake by continuously distributed mesophyll in leaves. J. Theor. Biol., 67, 471-88.

Parkhurst, D.F. (1978). The adaptive significance of stomatal occurrence on one or both surfaces of leaves. J. Ecol., 66, 367-83.

Parkhurst, D.F. & Loucks, O.L. (1972). Optimal leaf size in

relation to environment. J. Ecol., 60, 505-37.
Quarrie, S.A. & Jones, H.G. (1977). Effects of abscisic acid and water stress on development and morphology of wheat. J. Exp. Bot., 28, 192-203.
Rand, R.H. (1978). A theoretical analysis of CO_2 absorption in sun versus shade leaves. J. Biomech. Eng., 100, 20-4.
Sharma, K.D. & Sen, D.N. (1971). Sun and shade tolerance in the ecophysiology of Solanum surattense Burm. Zeit. Pflanzenphysiol., 64, 263-6.
Sheriff, D.W. (1979). Water vapour and heat transfer in leaves. Ann. Bot., 43, 157-71.
Shropshire, W. Jr. & Mohr, H. (1983). Photomorphogensis. Encyclopedia of Plant Physiology, New Series Vols. 16A, Berlin-Heidelberg-New York: Springer-Verlag.
Siwecki, R. & Kozlowski, T.T. (1973). Leaf anatomy and water relations of excised leaves of six Populus clones. Arboretum Kornickie 18, 83-105.
Smith, W.K. & Nobel, P.S. (1978). Influence of irradiation, soil water potential, and leaf temperature on leaf morphology of a desert broadleaf, Encelia farinosa Gray (Compositae). Am. J. Bot., 65, 429-32.
Ticha, I. (1982). Photosynthetic characteristics during ontogenesis of leaves. 7. Stomata density and sizes. Photosynthetica, 16, 375-471.
Van Volkenburgh, E. & Davies, W.J. (1983). Inhibition of light-stimulated leaf expansion by abscisic acid. J. Exp. Bot., 34, 835-45.
Walton, D.C. (1980). Biochemistry and physiology of abscisic acid. Ann. Rev. Plant. Physiol., 31, 453-89.
Willmer, C.M. (1983). Stomata. Harlow: Longmans.

TEMPERATURE AND LEAF GROWTH

C.K. Ong and C.K. Baker
Department of Physiology and Environmental Sciences, University of Nottingham, Sutton Bonnington, Leicester, LE12 5RD, United Kingdom.

INTRODUCTION

Knowledge of how temperature affects leaf growth is important for three main reasons; first, the importance of leaf area in determining the growth rate of crops, which is directly related to the radiation intercepted by the foliage during the growing season (Biscoe & Gallagher, 1977; Day, 1981). In agriculture, crop management usually affects leaf growth rather than the actual rate of photosynthesis per unit leaf area (Monteith & Elston, 1983). Second, the introduction of new crops into cooler or warmer regions is largely determined by a consideration of the influence of temperature (e.g. maize in Europe and North America, Bunting, 1976; banana in sub-tropical Australia, Turner & Lahav, 1983). Finally, even in regions where lack of rain is the main restraint on productivity, temperature is still an important factor determining leaf growth because water deficit and heat stress are commonly closely linked.

Much of the basic information on the response of leaf growth to single environment factors was obtained during the 60's when controlled-environment facilities became available (Friend, 1961; Kleinendorst & Brouwer, 1970) but it has proved difficult to extrapolate results obtained in controlled environments to the field because major features of the natural environment were sacrificed for the sake of control. Even now it is very expensive to simulate full sunlight (2000 E m^{-2} s^{-1}) on a working area large enough for an experiment which requires many plants to allow sequential harvesting. Disappointment with such attempts led many workers to make detailed and extensive measurements of the environment and crops in the field during the 70's (Biscoe et al., 1975).

The first major scientific meeting to consider the control of leaf growth was in 1956 (Milthorpe, 1956) and the latest in 1981 (Dale & Milthorpe, 1983). Two of the few advances during the intervening period was the identification of the shoot meristem as the site of temperature perception in cereals and grasses (Kleinendorst & Brouwer, 1970; Watts, 1972; Peacock, 1975), and the advent of reliable and robust electronic auxanometers (Williams & Biddiscombe, 1965; Gallagher et al., 1976). Monteith (1979) pointed out that the literature of the past 20 years has extended the evidence that crop plants respond to soil temperature but has not added much to our understanding of underlying mechanisms. Therefore, in this review we will attempt to define relevant concepts, establish principles and describe their application, and finally to indicate the areas where information is most needed.

GENERALISATIONS

Temperature affects the growth of a crop very soon after seed is sown, provided the soil is moist. Growth of leaf primordia which are partially or fully formed in the embryo of the seed will be less affected by soil temperature than leaves formed subsequently, because the major phase of cell division terminates by the time the leaf is between 10 - 20% of the final size especially in cereals (Dale, 1976). Cell division continues at a slow rate for the rest of the period of lamina expansion but the proportion of the time for each phase is species-dependent.

Fig. 1a illustrates the two ways of describing the response for rate of cell division or leaf expansion over a range of temperature. The line 1 represents a response with three sharply defined points - the so-called cardinal temperatures, base (T_b), optimum (T_o) and maximum (T_m). Examples of this first response are reported mainly for leaves of young seedlings, when light is non-limiting (Blacklow, 1972; Garcia-Huidobro et al., 1982) and of leaves

Fig. 1. General relationships between temperature (T) and leaf growth. Base, optimum and maximum temperatures are T_b, T_o and T_m respectively. Broken lines represent range where results are lacking. (Monteith, 1979.)

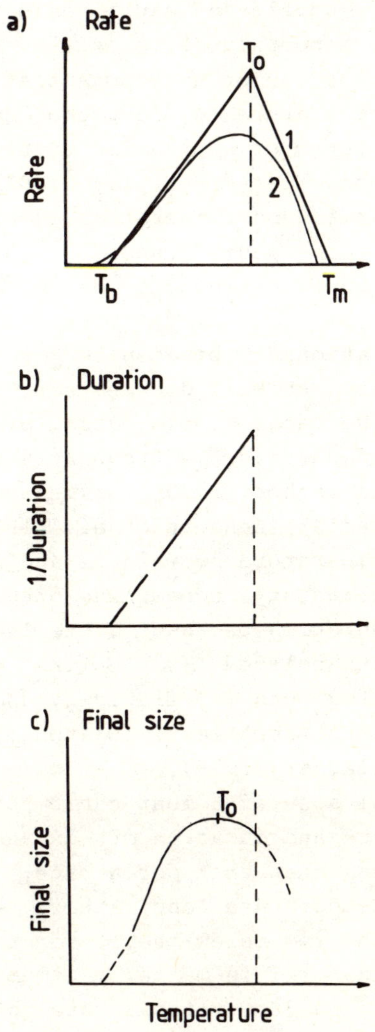

of plants grown in controlled-environment where evaporative demand is low (Fortainer, 1957; Turner & Lahav, 1983). The second response is a more general representation of many observations where curvature at high temperatures is common and may be a consequence of other environmental factors such as low light and low humidity or the result of a combination of a number of unrelated experiments. The degree of curvature at low temperatures is usually small but is the only reason for the use of an exponential function of Q_{10}. The effect of temperature on plant processes has therefore been analysed in terms of activation energy derived from the Arrhenius equation (Terry et al., 1983). In fact, the relation between rate and temperature is not exponential but linear over a temperature range of 10-15 K below the optimum, which explains why Q_{10} decreases with increasing temperature.

The relationship between temperature and duration of leaf growth is less well documented (Fig. 1b) although results for a wide range of biological processes show that the reciprocal of duration is frequently a linear function of temperature (Gallagher, 1979, for mean between 4 and 12°C for leaves of cereals; Dennett et al., 1979, for broad bean using maximum temperature between 12 and 27°C). There is still some uncertainty concerning the choice of base temperature for duration because of the lack of measurements close to the extrapolated T_b. Unlike growth rate, the duration of leaf growth is relatively unaffected by water stress, light or nutrient level (Elston et al., 1976; Auld et al., 1978; Gallagher, 1979).

The final size of a leaf can be described in terms of its mean rate and duration of expansion. If T_b for duration is less than that for a rate, final size will increase with temperature and attain a maximum at a temperature which is below the T_o for rate of expansion (Fig. 1c). For example, in maize, maximum size was achieved at 20°C but T_o is 30°C for rate of leaf expansion (Grobellar, 1963). In contrast, large differences in leaf size between leaves on the same plant are usually due to differences in rate rather than duration of expansion (Gallagher, 1979, for wheat; Dennett et al., 1979, for broad bean). Therefore analysis of final leaf size should consider both rate and duration of expansion.

Significance of Soil Temperature

Much of the relevant literature on the response of leaves to soil temperature is concerned with the response to cold of crop plants in temperate regions, where the distribution of crops and their appropriate planting dates are largely governed by sensitivity to low temperature (Cooper, 1973). Information on the response to high temperature is very limited even though yield, especially of tropical crops, is highly sensitive to high soil temperature (Lal, 1974; Harrison-Murray & Lal, 1979). Work on grasses and cereals shows that soil temperature has a major influence on leaf growth during the vegetative phase because the leaf meristem, the site of temperature perception, is situated below or close to the soil surface (Kleinendorst & Brouwer, 1970; Watts, 1972; Peacock, 1975). These workers were able to show that extension rate responds mainly to the temperature of the meristematic region at the base of the leaves, whereas changes in both air and root-zone temperatures are relatively unimportant. The apparent extension of maize leaves below 10°C in Watt's experiment was probably due to inadequate monitoring of the meristem temperature: later investigation showed that leaf extension of maize ceases at 10°C (Miedema & Sinnaeve, 1980).

As leaves emerge further and further away from the soil surface the influence of soil temperature declines. For example, in millet the meristem is clearly above the soil after 7 or 8 leaves have emerged, but soil temperature continues to exert some influence (Ong, 1983a). Both Watts (1972) and Adams & Thompson (1973) suggested that meristem temperature in maize and sorghum may be reduced by the cool water flowing through xylem vessels when transpiration is fast enough.

The practical importance of soil temperature as a

determinant of leaf growth, and hence yield, has been clearly demonstrated by the work of Cooper & Law (1977) on maize in the highlands of Kenya: when planting is delayed until after the onset of the rains and consequent soil cooling, yields are considerably reduced.

Day and Night Temperatures

There is still confusion in the literature concerning the method of analysis of the effects of temperature on leaf growth, which has an important bearing on the types of relationship derived. The method of calculating daily mean temperature is particularly important when night temperature falls well below the actual base temperature. For example, if the linear relationship established by constant temperature conditions is represented by two cardinal temperatures $T_b = 10^\circ C$ and $T_o = 30^\circ C$, the rate of leaf appearance at day/night (12 h) temperatures of 16/4 and 16/6°C are usually greater than rates predicted by a simple mean temperature calculated from both day and night values (Fig. 2). It can be shown that leaf growth rate at both 16/4 and 16/6°C is the same as that at 16/10°C, provided the low night temperature has no adverse effect on leaf growth during the higher day temperature. This type of result suggests that such temperatures (4 - 6°C below T_b) are equivalent to T_b *i.e.* the effective mean temperature is 13°C for both treatments and not 10 and 11°C respectively.

In contrast, Warrington & Kanemasu (1983) used a least-squares procedure to calculate the rates of leaf initiation and appearance for maize at low night temperatures (6 - 11°C) and predicted that T_b was approximately 4 and 2°C respectively for both processes. These minima are much lower than the value (10°C) reported for maize and other tropical cereals (Blacklow, 1972; Miedema & Sinnaeve, 1980; Ong, 1983a) and are closer to that reported for temperate crops (Angus *et al.*, 1980; Baker & Gallagher, 1983). Consequently, Warrington & Kanemasu (1983) concluded that the relationship between leaf growth of maize and temperature is curvilinear.

Fig. 2. Interpretation of leaf appearance rate when night temperature falls below T_b. Simple averages of day and night temperatures of 16/4 and 16/6°C regimes are represented by solid symbols. If temperature below T_b is equivalent to 10°C then effective mean temperature is 13°C (open symbol).

Fig. 3. Auxanometer measurements of leaf extension rate (R_e) in relation to air temperature during 9 - 11 April 1975 for leaf 7 in a field crop of winter wheat. Closed circles represent night-time measurements. The equation of the fitted line is, $y = 0.028x + 0.059$, where $r = 0.93$. (Gallagher, 1976.)

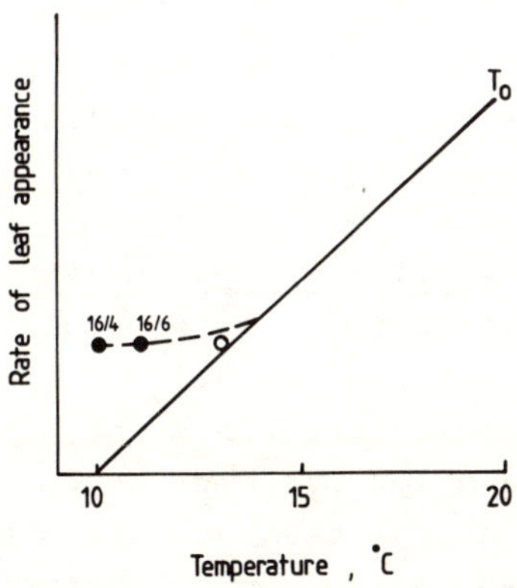

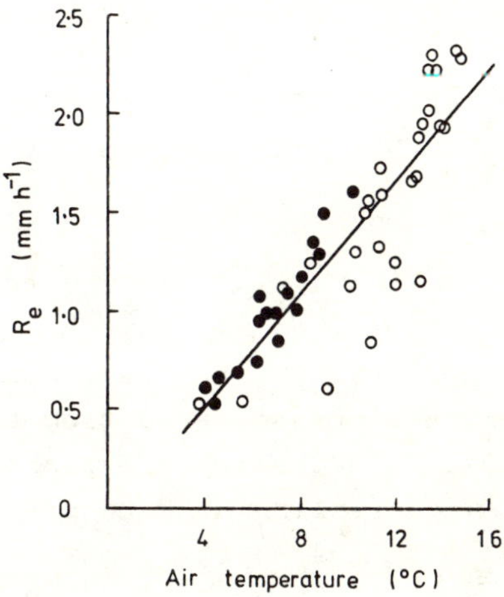

The significance of maximum and minimum temperatures (or day and night temperatures) for leaf growth is easily settled by using sensitive auxanometers to measure hourly extension rate, but surprisingly few attempts have been made, for example, to determine the actual values of T_b. Auxanometer measurements on winter wheat in the field (Gallagher, 1976; Baker, 1979) showed that the response of leaf extension to both day and night temperatures was fitted by the same straight line, with a T_b close to 0°C. Ong (1983b) showed that the extrapolated base temperature for leaf appearance rate is similar to the value determined for leaf extension rate with electronic auxanometers. It is more tedious to determine the significance of the appropriate temperature threshold for leaf duration. Nevertheless, it is curious that in an analysis for leaf duration and temperature Dennett et al., (1979) used maximum temperature although mean temperature gave an equally good correlation. The importance of maximum temperature for leaf growth is not well understood, except possibly where heat stress is common.

Crop physiologists have largely ignored the need to distinguish between the reversible and irreversible effects of temperature extremes on leaf growth. Work on processes, such as germination, has indicated that high temperature experienced by seeds for only a few hours can have an adverse effect on both the subsequent rate of germination and the final number that germinated (Garcia-Huidobro, 1982).

Thermal Time: Concept and Application

Thermal time, known elsewhere as "accumulated temperature" or "heat units", is not the amount of heat which is transferred from the environment to the plant, but time perceived by the plant as a function of temperature. It is generally used to predict or analyse the rate of plant development such as time to flowering or crop maturity (Cross & Zuber, 1972; Robertson, 1973) and has recently gained wider acceptance in analysing the influence of climatic factors and management on leaf growth (Gallagher, 1979; Baker et al., 1981; Kirby et al., 1982; Ong, 1983a; Squire & Ong, 1983; Milford, Pocock & Riley, personal communication).

The calculation of thermal time is based on the linear relationship between rate and temperature between T_b and T_o (Fig. 1a) although it can be easily modified to take into account temperature above T_o (Garcia-Huidobro et al., 1982). Linear relationships have been demonstrated for several crops (wheat and barley, Gallagher, 1979; broad bean, Dennett *et al.*, 1979; millet, Ong, 1983a; sugar beet, Milford & Riley, 1980; ryegrass, Keatinge *et al.*, 1979). The rate of leaf expansion R can be expressed as a thermally weighted rate, ρ defined by the equation

$$\rho = R / (T - T_b) \quad \text{where} \quad T_b < T < T_o \qquad (1)$$

where ρ is the reciprocal of the thermal time for a leaf to extend 1.0 mm, expressed as mm ($^{\circ}Ch^{-1}$) . Ong (1983b) has used this concept to establish the influence of ontogeny on the R of leaves. In general, T_b varies from 0 - 3°C for temperate crops and 8 - 13°C for tropical crops. For the duration between successive leaves or the duration of leaf growth (D) in d, a thermal duration (θ) can be expressed in °Cd by the equation

$$\theta = D(T - T_b) \quad \text{where} \quad T_b < T < T_o \qquad (2)$$

For temperatures which lie between T_o and T_m, θ can be calculated by the equation

$$\theta = D(T_m - T) \quad \text{where} \quad T_o < T < T_m \qquad (3)$$

Equation 3 has yet to be tested for leaf growth because the data are lacking.

There is still uncertainty concerning the choice of base temperature (T_b) because in all but a few instances the value of T_b is obtained by extrapolation, usually over several degrees. Factors which can influence the accuracy of the extrapolated T_b include sensitivity of leaf growth to other uncontrolled factors such as humidity, photoperiod, water deficit and the averaging of daily mean temperature. For millet the value of T_b for leaf initiation, appearance, expansion and canopy formation are conservative, 10 - 12°C (Ong, 1983a; Squire *et al.*, 1984). θ for leaf appearance of millet is relatively insensitive to humidity, photoperiod or mild water deficit (Ong, 1983a). In sugar beet, T_b for leaf appearance and expansion, without extrapolation, was reported to be 1°C and 3°C respectively (Milford, Pocock & Riley, personal communication).

Nevertheless despite uncertainties over T_b thermal time is still the most useful and meaningful method of analysis to separate the effect of temperature on leaf growth in the field, where temperature and other factors vary simultaneously. For example, Gallagher and his co-workers have shown that the production of leaf and spikelet primordia, leaf appearance, lamina expansion and the duration of leaf growth of field grown wheat and barley can all be described in terms of thermal time (Gallagher, 1979; Baker & Gallagher, 1983).

Leaf Initiation and Appearance

The initiation of leaf primordia of a species can be governed by both temperature and photoperiod (Rawson, 1971; Halse & Weir, 1974) and there are only a few attempts to interpret the effects of temperature on the rate of primordium initiation in the field (Baker & Gallagher, 1983). For example, θ for the initiation of leaf primordia was reported to be 63°Cd for winter wheat (Gallagher, 1979) and 21°Cd for spring wheat (Kirby, 1974). Because initiation is also influenced by photoperiod these values of θ, measured at 8 and 14.5 daylength, do not represent an inherent difference between the varieties. Baker & Gallagher (1983) assumed a multiplicative interaction between temperature and photoperiod for winter wheat and used a linear equation, similar to equation 1, to examine the combined effect of both factors. In species which are less sensitive to daylength θ for leaf initiation is relatively constant for a wide range of environmental conditions e.g. 26°Cd for pearl millet (Ong, 1983a) and 17°Cd for maize (Warrington & Kanemasu, 1983). In millet θ is similar for leaf initiation and appearance but reports for other species generally give a higher value for leaf appearance (e.g. 111°Cd for winter wheat, Gallagher, 1979, and 45°Cd for maize, Warrington & Kanemasu, 1983).

The rate of leaf appearance is generally less sensitive to other environmental conditions than is leaf expansion, especially in cereals, where the unemerged leaves are close to the ground and well protected from water deficit (Fig. 4 and Tables 1 and 2). Analysis of the stage of leaf appearance in terms of thermal time from time of emergence is particularly useful to illustrate the response to temperature (Fig. 5). Baker et al. (1980) and Kirby et al. (1983) found an effect of a rate of change of daylength on leaf appearance rate in the field but failed to detect a similar response in controlled environments.

Fig. 4. Increase in leaf number (N) and leaf area index (LAI) on the main stem of pearl millet under three water regimes: wet (●), intermediate (○) and dry (□).

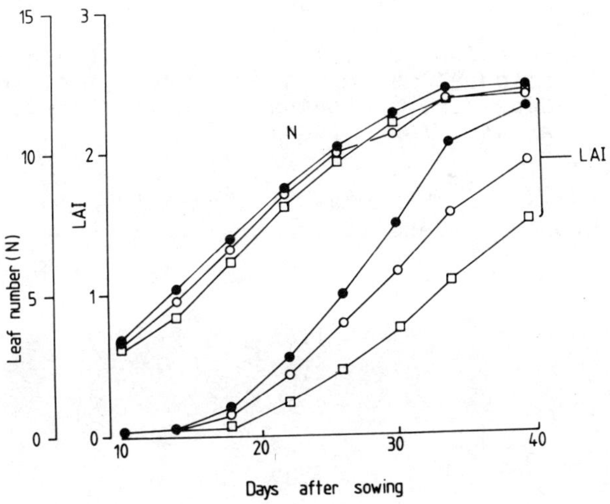

Table 1. Comparison of thermal duration θ, for leaf appearance of pearl millet at various irradiances and daylengths. All values are calculated at weekly intervals for leaves 2 to 9 during 1978 and using a T_b of 12.4°C (Ong, 1983a).

Mean soil temperature at 5cm depth (°C)	θ (°Cd per leaf)	Mean daily irradiance ($MJ\ m^{-2}$)	Mean daylength (including civil twilight) (h)
16.9	25.0	4.4	14.2
19.1	24.8	4.9	15.7
21.2	23.8	9.0	17.4
24.2	24.5	6.8	18.0
27.4	22.7	6.5	18.5
25.6	26.4	9.1	18.5

Mean value for θ = 23.9 ± 1.3°Cd

Table 2. Comparison of thermal duration, θ, for leaf appearance of pearl millet at different saturation deficits. θ is calculated for each glasshouse with T_b = 12.0°C and mean meristem temperature. Standard deviators are given in brackets. (Ong, 1983a.)

Mean air temperature (°C)	Mean maximum saturation deficit (kPa)	θ (°Cd per leaf)
19.6	1.55	24.6 (3.7)
22.7	2.27	25.9 (5.6)
25.8	2.33	28.3 (2.4)
28.5	2.98	36.3* (0.2)
30.4	3.07	26.5 (2.7)

* Meristem temperature exceeded 37°C on four days

Fig. 5. (A) Course of leaf appearance at four sowing dates. Dates of sowing: (●) 20 March, (○) 10 April, (■) 19 May and (□) 13 June. (B) Rate of leaf appearance in relation to soil temperature (5.0 cm). $1/t = (T - 12.4)/23.8$, $r^2 = 0.83$. (C) Leaf appearance stage in relation to thermal time after sowing. Symbols for dates of sowing as in Fig. 5A. $T = 12.4°C$. (Ong, 1983a).

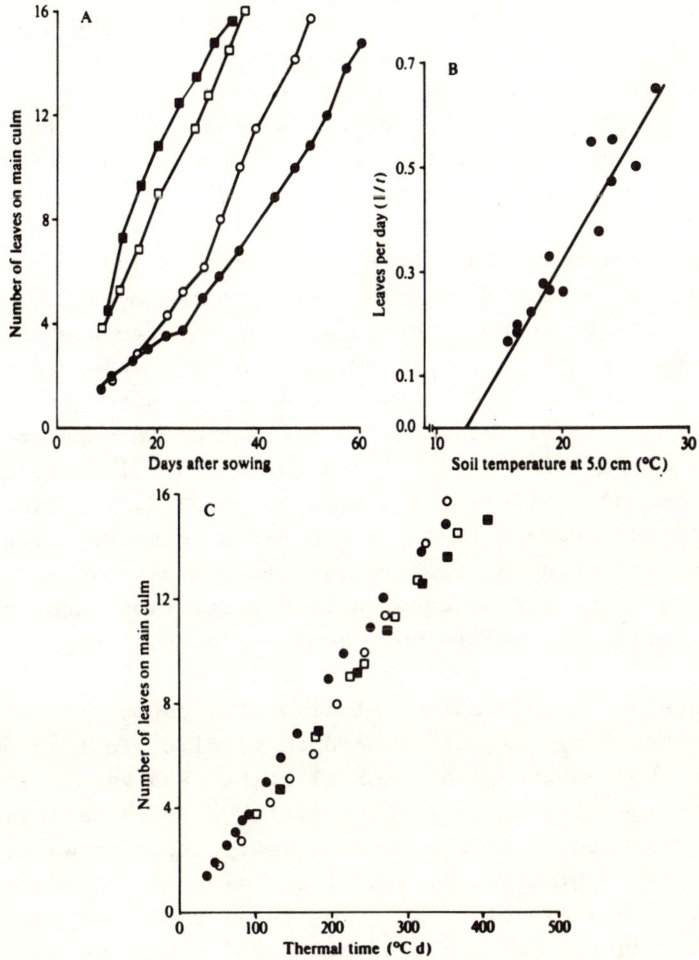

Rate and Duration of Leaf Expansion

The final size of a leaf is the product of the rate and duration of expansion. Leaf size is influenced by the initial number of cells in the primordium, the rate and duration of cell division, and the size of the mature cells, but cell number is generally considered to be the major determinant of leaf size in many species (Abbe et al., 1941; Ashby & Wangermann, 1950; Cannell, 1978). Evidence of Williams (1960) and Dale (1976) suggests that though division and extension are concurrent throughout most of the growth of the leaf it is the extension of cells that must contribute to increase in leaf size.

There has been no systematic study to show how temperatures during the early life of a leaf influence its subsequent expansion and final size. Dale (1982) compared the effect of two temperatures (15 and 20°C) on the leaf growth of wheat in growth rooms and concluded that neither cell number nor final size was affected by thermal history. The lack of response could have arisen because most of the leaves studied were already initiated in the embryo of the seed, or because only two temperatures were used: Friend & Pomeroy (1970) found strong effects of temperature on size for comparable wheat leaves grown at 10 - 30°C. Field studies of leaf expansion of sugar beet during five seasons (Milford, Pocock & Riley, personal communication) suggest that the influence of temperature (and nitrogen) is greatest during the period between leaf initiation and unfolding, when cell division is most active (Terry, 1968).

Temperature greatly influences the rate of leaf expansion in chronological time (R_e) especially for leaves in the field so it is hard to resolve leaf growth data without recourse to thermal time analysis. From his extensive investigations on winter cereals Gallagher (1979) demonstrated that the rate of leaf extension was steady at constant temperature, and cited evidence (Richards, 1969) that this would be expected when cell extension predominates. For any leaf of wheat or barley, he found that about the first 10% of final length grew exponentially but that the remaining 90% of growth was linear when plotted against thermal time calculated above 0°C. More recently Ong (1982b) has demonstrated that millet, a tropical cereal, behaves similarly and that the base temperature for R_e is 10°C. In sugar beet R_e is linearly related to mean temperature in the linear growth phase but length and width growth alternate during each 24 h cycle in patterns not clearly related to temperature or the day/night sequence.

A systematic difference in the size of successive leaves has been reported from several species of the Gramineae (Borrill, 1959, rye-grass; Kirby, 1973, barley; Gallagher, 1976, wheat). Similar ontogenetic size gradients are also found in dicotyledons, *e.g.* sunflower (Rawson & Hindmarsh, 1982) beans (Dennett *et al.*, 1979), and sugar beet (Milford, Pocock & Riley, personal communication). For temperate cereals, thermal time analysis shows that ontogenetic leaf size differences are associated with successively higher values of expansion per unit of thermal time (ρ) with leaf position up the stem. If data for leaves from different positions are mixed in analysis, a curvilinear response of R_e to mean temperature results (*e.g.* Peacock, 1975; Williams & Biddiscombe, 1965) and this obscures the true linear response for leaves from the same position. Curiously, this method of analysis can sometimes yield a straight-line response, possibly if large enough numbers of R measurements are combined to obtain means (Keatinge *et al.*, 1979). Ontogenetic differences in final leaf size apparently originate because during leaf initiation the stem apex is expanding, so that each primordium is slightly larger than its predecessor and contains more cells from the start (Abbe *et al.*, 1941).

The duration of leaf growth, D_e, has been shown to be directly related to growth temperature in wheat and beans respectively. Gallagher (1976) and Dennett *et al.*, (1979) found that during the linear growth phase D_e was a linear function of temperature. This implies that at a constant temperature D_e is similar for all leaves and that ontogenetic differences in leaf size must therefore be due to differences in ρ. Growth duration was constant at constant temperature in tobacco (Wolf, 1947), where D_e did not change with leaf position despite an ontogenetic size gradient. In contrast, for sunflowers grown in controlled environments at constant temperature by Rawson & Hindmarsh (1982) showed increased D_e (and R_e) with higher leaf position. With increasing temperature sunflower leaf size at a given position decreased, because - unusually - R_e did not increase; R_e was apparently limited by lack of assimilate, resulting from poor light penetration into the stand of plants established in the growth-cabinet. Results like these emphasise the pitfalls of artificial environments and the desirability of conducting field studies from which the data can be interpreted using thermal time analysis, and the importance of ensuring that the effects of temperature are not confounded by other factors.

The basis of the relationship between temperature and D_e is not clear. In a determinate cereal, such as wheat, the ear cannot emerge until the series of leaves subtending it has expanded. It is tempting to speculate that in such a case the constancy of thermal duration for leaf expansion represents division of the total thermal time to ear emergence into equal-length segments whose number depends on how many leaves have to appear and unfold. For beans, Auld _et al_. (1978) suggested that D_e may well be linked to the rate of development of young competing leaves, which depends strongly on temperature.

Morphology

Temperature can affect leaf morphology, though usually more weakly than it affects expansion. Turner & Lahav (1983) found that banana leaves were thinner at higher temperatures. In several grasses leaves tended to be thinner with increasing night temperature but thicker for a corresponding increase of day temperature (Ivory & Whiteman, 1978). Friend _et al_. (1962) showed that though wheat leaves became longer with increasing temperature they became narrower and thinner. The effect of higher temperature was to increase the number and length of epidermal cells; mesophyll cells became not only longer but also more lobed.

Canopy Formation

Temperature has a strong influence on the rate of canopy formation and the duration of the canopy in a determinate crop. This review will concentrate on the former although the negative relation between the duration of the canopy and grain yield is more important for cereal production in Britain (Monteith, 1981). The rate of canopy formation depends on the rate and duration of leaf growth as well as the number of branches or tillers per unit area of ground. Although there are numerous studies of how temperature influences growth of single leaves or plants (Mitchell, 1956; Baker & Jung, 1968; Ivory & Whiteman, 1978) there is a dearth of information on how temperature affects canopy formation.

Squire et al. (1984) compared the rate of canopy formation, light interception and dry matter production of five stands of pearl millet grown at mean air temperature of 19, 22, 25, 28 and 31°C in glasshousses. The rate of increase of leaf area index (LAI) was linearly related to temperature above an extrapolated base of 10.0 ± 0.5°C (Fig. 6). The shape of this relation is similar to that for the extension of individual leaves of millet with a T_b of 10°C without extrapolation (Ong, 1983b). A thermal time of 300°Cd was required from sowing to the start of rapid expansion and 500°Cd to LAI = 5. The total is consistent with the sum of thermal times of the various processes in canopy formation measured individually. 40°Cd was required for seedling emergence, 325°Cd for the initiation of 13 leaves and 80°Cd for the expansion of the flag leaf. A further 50°Cd was required to account for the slowest plants in the stand.

Temperature had little influence on the maximum LAI reached in the above experiment because temperature during the leaf initiation phase was the same in all treatments. More generally, however, temperature during the early vegetative growth may influence canopy size through its effect on the numbers of leaves per tiller and of tillers per plant (Pearson, 1975, for millet).

Fig. 6. The relation between temperature and the rate of increase of LAI of pearl millet stands during the linear phase of canopy formation. Mean air temperatures are indicated. Omitting the value for 31°C, the regression is $y = 0.021\ (T - 10.0)$, where $r = 0.96$. (Squire et al., 1984.)

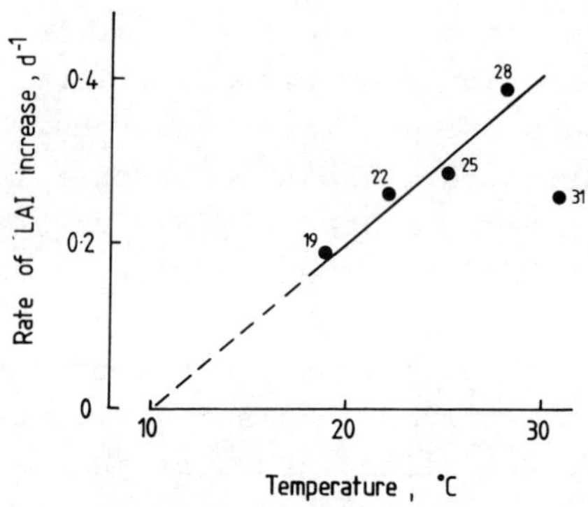

Studies of the final leaf number in twelve sorghum varieties grown at five mean air temperatures showed that most cultivars produced more leaves (3-10) as temperatures increased from 21 to 30°C (Quinby et al., 1973). Increase in leaf number can be explained by the faster rate of leaf initiation because the duration of vegetative phase is relatively less sensitive to temperature than leaf initiation. At the lowest temperature (13.5°C) leaf number was generally higher (0.5 - 5) even though the rate of leaf initiation was much lower than at 21°C and the increased leaf number was due to a longer (three times) vegetative phase.

Because leaf area increase is so temperature-dependent, in the field, the fluctuating temperatures characteristically experienced can result in a confusing picture of events, as in the increase of LAI with time for sugar beet crops (Milford, Pocock & Riley, personal communication; Fig. 7a). With thermal time analysis, however, the trends observed in such data are clearly revealed (Fig. 7b).

Varietal Differences

In temperate regions, the slow leaf growth of many crops during early spring is regarded as a major limitation to crop productivity (Scott *et al.*, 1973). In spring, when solar radiation is high temperatures are relatively low and consequently leaf development is slow. It has therefore been argued that it would be worthwhile to develop varieties which are capable of producing a faster early leaf cover at low temperatures (Monteith & Elston, 1971; Fick *et al.*, 1975). Comparison of nine sugar beet varieties at constant temperatures of 7, 11, 15 and 20°C showed large genetic differences in the rate of leaf growth were more clearly related to variation in the rate of leaf expansion than to the rate of leaf appearance or the duration of leaf expansion, which were relatively constant between varieties (Milford & Riley, 1980).

There are few attempts to select for varieties which could grow earlier in the season *i.e.* genotypic differences in T_b although it is well known that genetic differences in response to low temperature, *e.g.* in frost sensitivity, do exist. Miedema & Sinnaeve (1980) concluded that the growth of maize ceased at 10°C because it was limited by cell division and/or expansion but their search for genotypic differences in T_b was unsuccessful (personal communication). In contrast, unpublished data of Mohammed showed that T_b of contrasting millet varieties ranged from 8 - 13°C but differences in leaf expansion rate were still the major cause of large variation in leaf growth. In the semi-arid tropics differences in leaf growth at high temperatures are important but there is no report of significant differences in T_m between varieties. Selection

Fig. 7. The increase of LAI for two pearl millet stands in (a) chronological and (b) thermal time. Mean temperature was 25°C until 22d, then either 28 (●) or 19°C (○). (Squire et al., 1984.)

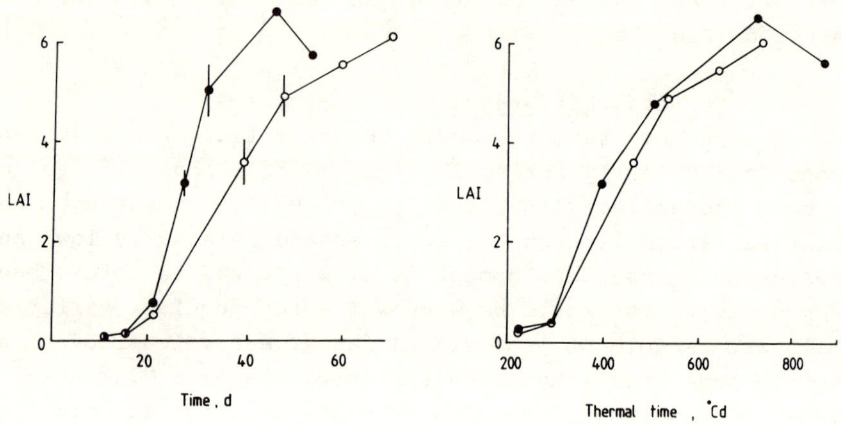

for higher leaf expansion rate in sorghum at high temperature is already an important selection criterion of the International Crop Research Institute for the Semi-Arid Tropics (Peacock, 1982).

CONCLUSIONS

In this review we have described the several responses of leaf growth to a wide range of temperatures, the concept of thermal time and assessed its usefulness, and highlighted the difficulties of examining leaf growth in the field. The effect of temperature on other aspects of leaf physiology such as leaf persistence, tillering, functioning and death, are not covered here.

The following list summarises priorities for future research.

1. Leaf initiation. How important is thermal history and cell division and cell number during leaf initiation on subsequent leaf expansion and size? Field results indicate that the duration of leaf expansion is less variable than expansion rate and the latter is the major

determinant of the rate of formation and size of canopies.

2. Leaf appearance. There is still considerable uncertainty concerning the interaction between temperature and the rate of daylength change at seedling emergence. Unpublished results indicate that daylength after the 2-3 leaf stage does not appear to influence the value of θ of leaf appearance rate.

3. Leaf duration. The influence of supraoptimal temperatures on the duration of leaf expansion is not known, although it is generally recognised as a major limitation to leaf growth in the semi-arid tropics.

4. Leaf size. The influence of extremes of temperature on leaf size, as distinct from the indirect response associated with vernalisation or flowering stimulus, is not well understood.

5. Genotype differences. Information on genotype differences in the value of T_b and T_o is still lacking although selection for earlier growing varieties could improve crop performance.

Progress in the research in these areas is needed to anable a meaningful synthesis or modelling (see Gallagher, this volume) of the various aspects of leaf growth and their physiological activity and finally, to use such knowledge to explain the variation in crop performance from season to season and from site to site.

REFERENCES

Abbe, E.C., Randolph, L.F. & Einset, J. (1941). The developmental relationship between shoot apex and growth pattern of leaf blade in diploid maize. Amer. J. Bot., 28, 778-84.

Adams, J.E. & Thompson, D.O. (1973). Soil temperature reduction during pollination and grain formation of corn and grain sorghum. Agron. J., 65, 60-63.

Angus, J.F., Cunningham, R.B., Moncur, M.W. & MacKenzie, D.H. (1981). Phasic development in field crops. 1. Thermal response in the seedling phase. Field crop Res., 3, 365-78.

Ashby, E. & Wangerman, E. (1950). Studies on the morphogenesis of leaves. IV. Further observations on area, cell size and cell number of leaves of Ipomoea in relation to their position on the shoot.

New Phytol., 49, 23-35.

Auld, B.A., Dennett, M.D. & Elston, J. (1978). The effect of temperature changes on the expansion of individual leaves of *Vicia faba* L. Ann. Bot., 42, 877-88.

Baker, B.S. & Junge, G.A. (1968). Effect of environmental conditions on the growth of four perennial grasses. 1. Response to controlled temperatures. Agron. J., 60, 155-8.

Baker, C.K. (1979). The environmental control of development in winter wheat. Ph.D. thesis, University of Nottingham.

Baker, C.K. & Gallagher, J.N. (1983). The development of winter wheat in the field. 2. The control of primordium initiation rate by temperature and photoperiod. J. Agric. Sci., Camb., 101, 337-44.

Baker, C.K., Gallagher, J.N. & Monteith, J.L. (1980). Daylength change and leaf appearance in winter wheat. Plant, Cell and Environ., 3, 285-87.

Blacklow, W.M. (1972). The influence of temperature in germination and elongation of the radicle and shoot of corn (Zea mays L.). Crop Sci., 12, 647-50.

Bunting, G.S. (1976). Accumulated temperature and maize development in England. J. Agric. Sci., Camb., 87, 577-583.

Biscoe, P.V., Clark, J.A., Gregson, K., McGowan, M., Monteith, J.L. & Scott, R.K. (1975). Barley and its environment. 1. Theory and practice. J. Appl. Ecol., 12, 227-57.

Biscoe, P.V. & Gallagher, J.N. (1977). Weather, dry matter production and yield. *In* "Environment Effects on Crop Physiology". Eds. J.J. Landsberg & C.V. Cutting, pp. 75-100. London: Academic Press.

Cooper, A.J. (1973). Root temperature and plant growth. Commw. B. Hort. Plantn. Crops.

Cooper, J.P. & Tainton, N.M. (1968). Light and temperature requirements for the growth of tropical and temperate grasses. Herb. Abstr., 38, 167-76.

Cooper, P.J.M. & Law, R. (1977). Soil temperature and its association with maize yield variations on the Highlands of Kenya. J. Agric. Sci., Camb., 89, 355-363.

Cannell, M.G.R. (1978). Analysis of shoot apical growth of *Picea sitchensis* seedlings. Ann. Bot., 42, 1291-1303.

Cross, H.Z. & Zuber, M.J. (1972). Prediction of flowering dates in maize based on different methods of estimating thermal units. Agron. J., 64, 351-55.

Dale, J.E. (1976). Cell division in leaves. *In* "Cell division in Higher Plants". Ed. M.M. Yeoman, pp. 315-45. London: Academic Press.

Dale, J.E. (1982). Some effects of temperature and irradiance on growth of the first four leaves of

wheat, Tricicum aestivium. Ann. Bot., 50, 851-58.
Dale, J.E. & Milthorpe, J.L. (1983). General features of the production and growth of leaves. In :The Growth and Functioning of Leaves". Eds. J.E. Dale & F.L. Milthorpe, pp. 151 178. Cambridge: Cambridge University Press.
Day, W. (1981). Water stress and crop growth. In "Physiological Processes Limiting Plant Productivity", ed. C. Johnson. Proceedings of 30th Easter School, University of Nottingham. London: Butterworths.
Dennett, M.D., Elston, J. & Milford, J.R. (1979). The effect of temperature on the growth of individual leaves of Vicia faba L. in the field. Ann. Bot., 197-208.
Elston, J., Karamanos, A.J., Kassam, A.H. & Wadsworth, R.M. (1976). The water relations of the field bean crop. Phil. Trans. R. Soc. Lond. B., 273, 581-91.
Fortainer, E.J. (1957). Control of flowering in Arachis hypogaea L. Meded. Landbhoogesch, Wageningen. Ph.D. Thesis.
Friend, D.J.C., Helson, V.A. & Fisher, J.A. (1962). Leaf growth in Marquis wheat as regulated by temperature, light intensity and daylength. Can. J. Bot., 40, 1299-1311.
Friend, D.J.C. & Pomeroy, M.E. (1970). Changes in cell size and number associated with the effects of light intensity and temperature on the leaf morphology of wheat. Can. J. Bot., 48, 85-90.
Gallagher, J.N. (1979). Field studies of cereal leaf growth. I. Initiation and expansion in relation to temperature and ontogeny. J. Exp. Bot., 30, 625-36.
Gallagher, J.N., Biscoe, P.V. & Saffell, R.A. (1976). A sensitive auxanometer for field use. J. Exp. Bot., 27, 704-16.
Garcia-Huidobro, J. (1982). Control by temperature and soil water of germination and emergence in pearl millet. Ph.D. Thesis, University of Nottingham.
Garcia-Huidobro, J., Monteith, J.L. & Squire, G.R. (1982). Time, temperature and germination of pearl millet (Pennisetum typhoides S. & H.). I. Constant temperature. J. Exp. Bot., 33, 287-95.
Grobelaar, W.P. (1963). Response of young maize plants to root temperature. Med. Landbouwhog, Wageningen, 63, 1-71.
Harrison-Murray, R.S. & Lal, R. (1979). High soil temperature and the response of maize to mulching in the lowland humid tropics. In "Soil Physical properties and Crop Production in the Tropics". Eds. R. Lal & D.J. Greenland, pp. 285-304. London: Wiley.
Ivory, D.A. & Whiteman, P.C. (1978). Effect of temperature on growth of five subtropical grasses. 1. Effect of

day and night temperature on growth and morphological development. Aust. J. Plant Physiol., 5, 131-48.

Keatinge, J.D.H., Stewart, R.H. & Garrett, M.K. (1979). The influence of temperature and soil water potential on the leaf extension rate of perennial ryegrass in Northern Ireland. J. Agric. Sci., Camb., 92, 175-83.

Kirby, E.J.M. (1974). Ear development in spring wheat. J. Agric. Sci., Camb., 82, 437-47.

Kirby, E.J.M., Appleyard, M. & Fellowes (1982). Effect of sowing date on the temperature response of leaf emergence and leaf size in barley. Plant, Cell and Environ., 5, 477-84.

Kleinendorst, R. & Brouwer, R. (1970). The effect of temperature of the root medium and of the growing part of the shoot on growth, water content and sugar content of maize leaves. Neth. J. Agric. Sci., 18, 140-48.

Lal, R. (1974). Soil temperature relations in tropical Africa and their effect on crop yield. International expert consultation on the use of improved technology for food production in rainfall areas of tropical Asia, FAO.

Miedema, P. & Sinnaeve, J. (1980). Photosynthesis and respiration of maize seedlings at suboptimal temperatures. J. Exp. Bot., 31, 813-19.

Milford, G.F.J. & Riley, J. (1980). The effects of temperature on leaf growth of sugar beet varieties. Ann. Appl. Biol., 94, 431-43.

Milthorpe, F.L. (1956). The growth of leaves. Proceedings of 3rd Easter School in Agricultural Science, University of Nottingham. London: Butterworths.

Mitchell, K.J. (1956). Growth of pasture species under controlled environments. 1. Growth at various levels of constant temperature. N.Z. J. Sci. Tech. Sect. A., 38, 203-16.

Monteith, J.L. (1979). Soil temperature and crop growth in the tropics. In "Soil Physical Properties and Crop Production in the Tropics", eds. R. Lal & D.J. Greenland, pp. 249-62. New York: Wiley.

Monteith, J.L. (1981). Climatic variation and the growth of crops. Quart. J.R. Met. Soc., 107, 749-74.

Monteith, J.L. & Elston, J. (1983). Performance and productivity of foliage in the field. In "The growth and functioning of leaves", eds. J.E. Dale & F.L. Milthorpe, pp. 499-518. Cambridge: Cambridge University Press.

Ong, C.K. (1983a). Response to temperature in a stand of pearl millet (Pennisetum typhoides S. & H.) 1. Vegetative development. J. Exp. Bot., 34, 322-36.

Ong, C.K. (1983b). Response to temperature in a stand of

pearl millet (Pennisetum typhoides S. & H.) 4. Extension of individual leaves. J. Exp. Bot., 34, 1731-39.

Peacock, J.M. (1975). Temperature and leaf growth in Lolium perenne II. The site of temperature perception. J. Appl. Ecol., 12, 115-23.

Peacock, J.M. (1982). Response and tolerances of sorghum to temperature stress. Sorghum in the Eighties. A symposium of ICRISAT, Hyderabad, India.

Quinby, J.R., Hesketh, J.D. & Voigt, R.L. (1973). Influence of temperature and photoperiod on floral initiation and leaf number in sorghum. Crop Sci., 13, 243-46.

Rawson, H.M. & Hindmarsh, J.H. (1982). Effects of temperature on leaf expansion in sunflower. Aust. J. Plant Physiol., 9 201-19.

Richards, F.J. (1969). The quantitative analysis of growth. In "Plant Physiology", Vol. VA, ed. F.C. Steward, pp. 3-76. New York: Academic Press.

Robertson, G.W. (1973). Development of simplified agroclimatic procedures for assessing temperature effects. In "Plant Response to Climatic Factors", ed. R.O. Slatyer. Proc. Uppsala Symp. 1970. UNESCO, Ecol. Conserv. 5.

Scott, R.K., English, S.D., Woad, D.W. & Unsworth, M.H. (1973). The yield of sugar beet in relation to weather and length of growing season. J. Agric. Sci., Camb., 81 339-47.

Squire, G.R., Marshall, B., Terry, A.C. & Monteith, J.L. (1984). Response to temperature in a stand of pearl millet. 5. Light interception and dry matter production. J. Exp. Bot., 35, 599-610.

Squire, G.R. & Ong, C.K. (1983). Response to saturation deficit of leaf extension in a stand of pearl millet (Pennisetum typhoides S. & H.). 1. Interaction with temperature. J. Exp. Bot., 34, 846-55.

Terry, N. (1968). Developmental physiology of sugar beet. 1. The influence of light and temperature on growth. J. Exp. Bot., 19, 795-811.

Terry, N., Waldron, L.J. & Taylor, S.E. (1983). Environmental influences on leaf expansion. In "The Growth and Functioning of Leaves", eds. J.E. Dale and F.L. Milthorpe, pp. 179-206. Cambridge: Cambridge University Press.

Turner, D.W. & Lahav, E. (1983). The growth of banana plants in relation to temperature. Aust. J. Plant Physiol., 10, 43-53.

Watts, W.R. (1972). Leaf extension in Zea mays. II. Leaf extension in response to independent variation of the temperature of the apical meristem, of the air around the leaves, and of the root zone. J. Exp. Bot., 23, 713-21.

Warrington, I.J. & Kanemasu, E.T. (1983). Corn growth response to temperature and photoperiod. II. Leaf initiation and leaf appearance rates. Agron. J., 75, 755-61.

Williams, C.N. & Biddiscombe, E.R. (1965). Extension growth of grass tillers in the field. Aust. J. Agric. Res., 16, 14-22.

Williams, R.F. (1960). The physiology of growth in the wheat plant. 1. Seedling growth and the pattern of growth at the shoot apex. Aust. J. Biol. Sci., 13, 401-428.

DIFFERENCES IN MORPHOLOGY AND CHEMICAL COMPOSITION OF LEAVES GROWN AT DIFFERENT LIGHT INTENSITIES AND QUALITIES

Hartmut K. Lichtenthaler
Botanical Institute, University of Karlsruhe,
Kaiserstrasse 12, D-7500 Karlsruhe,
Federal Republic of Germany.

INTRODUCTION

Growth and development of plants and leaves are strongly influenced by light. This can best be visualised, when one compares the appearance of dark-grown, etiolated seedlings with that of light-grown plants. Dicotyledonous plants possess in the light shorter hypocotyls and epicotyls than dark-grown seedlings and also exhibit a leaf area growth (cotyledons, primary leaves) and a fully functional photosynthetic apparatus, which are missing in the etiolated seedlings. Similar light-induced growth responses are also found during the development of monocotyledonous plants. This light-dependent morphological development of seedlings, known as photomorphogenesis, is induced and regulated in many parameters by the active phytochrome, Pfr.

Photomorphogenesis of plants, and in particular the development of leaves, greatly depends on the amounts of incident light available during growth. In full sun light and under high light growth conditions (high quanta fluence rates) plants exhibit a strong light growth response (e.g. sun leaves of trees, high-light plants), which is quite different from the weak-light growth response seen in shade leaves and in low-light plants. These two different growth responses demonstrate the ability of plants to adapt to high and low light quanta fluence rates. Both adaptations are associated with special morphological, physiological and ultrastructural changes of the leaves and of their photosynthetic apparatus. These basic growth responses of plants are best seen at the two extremes, high light (full sunlight) and low light (shade), but depending on the intensity of the incident light, all transitional stages between the two extremes can be found.

The differences in morphology, chemical composition and physiological function between sun and shade leaves as well as high light and low light leaves will be treated in this chapter. The effects of light quality, in particular blue light and red light, will briefly be considered. It is also shown that chemicals such as herbicides and growth regulators can interfere with the light-adaptation response.

SUN AND SHADE LEAVES

The sun exposed leaves (sun leaves) of many broadleaf trees in general exhibit a smaller leaf area than the shade leaves from the partly or fully shaded inner part of the tree. This is well seen in the sun and shade leaves of the beech tree which exhibits a large light adaptation morphological response (Fig. 1). For single growing trees there exists a gradient in the light adaptation response between extreme sun leaves on the tree top and sun leaves on the sun exposed outer and lower parts of the tree, as well as between shade leaves and the extreme shade leaves of the inner tree canopy, which rarely see direct sunlight. This is documented by decreasing values for the ratio chlorophyll *a*/*b* from extreme sun (4-5) to extreme shade leaves (2.8-2.6). Blue shade leaves, which grow on the north side of trees and only receive only receive the blue skylight, show a light adaptation response, which lies in between that of sun and shade leaves.

Though the average leaf area of usual sun leaf of the beech is smaller (25 cm^2) than that of the usual shade leaf (38 cm^2), the average number of lateral leaf veins turning off on one side from the mid-vein is 9 in both cases (Table 1). This indicates that the leaf extension growth between the lateral veins is much greater in shade than in sun leaves. Correspondingly the frequency of stomata is considerably higher in sun (*ca.* 215 mm^{-2}) than in shade leaves (*ca.* 150 mm^{-2}). However, the total number of stomata per average sun and shade leaf is similar (Table 1). Between

Fig. 1. Differences in leaf area of sun (left) and shade (right) Leaves of the beech (Fagus sylvatica L.) View (a) on lower and (b) on upper surface.

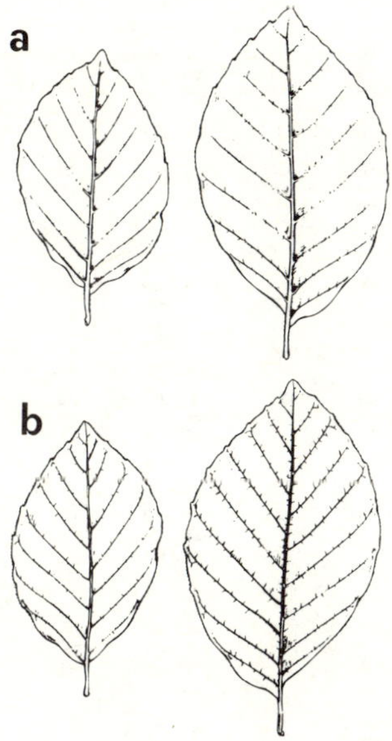

sun and extreme sun leaves or the shade and extreme shade leaves, there are no differences in the stomatal density. Thus stomatal density would appear to be a good parameter to differentiate between sun and shade leaves. Since the extreme sun or shade leaves exhibit a much smaller or larger leaf area respectively than the normal sun and shade leaves, the two extreme leaf types differ greatly in their total number of stomata (Table 1).

Sun leaves of trees are as a rule significantly thicker than the shade leaves. The sun leaves of the beech contain two rows of palisade parenchyma cells, compared to one row in shade leaves, as can be seen on leaf sections and freeze-fracture micrographs (Fig. 2 and 3). Correspondingly the total palisade parenchyma layer of sun leaves is thicker than the spongy parenchyma (Table 1). In contrast, the spongy parenchyma layer of shade leaves, though thinner than

Table 1. Some characteristics of fully developed, normal and extreme sun and shade leaves of Fagus sylvatica. Mean values and standard errors from 20 leaves harvested in July.

	extreme sun leaf	sun leaf	shade leaf	extreme shade leaf
ratio chlorophyll a/b	4.0 - 5.0	3.3 - 4.2	2.6 - 2.9	2.6 - 2.8
average leaf area (cm^2)	15	25	38	55
number of lateral veins (per half-leaf)	6 - 7	8 - 9	9 - 10	10 - 11
stomatal density (mm^{-2})	217 ± 13	214 ± 26	144 ± 11	151 ± 10
total stomata per leaf ($x10^3$)	325 (±6%)	535 (±12%)	547 (±8%)	830 (±7%)
total leaf thickness (μm)	201	171	92	93
tissue thickness (μm)				
upper epidermis	15±1	13±1	11±1	11±1
upper palisade cell	60±5	51±7		
lower palisade cell	43±6	32±4		
total palisade parenchyma	103±8	83±9	34±7	29±5
spongy parenchyma	69±10	63±11	36±6	42±6
lower epidermis	14±1	12±1	11±1	11±1
$\frac{\text{palisade parenchyma}}{\text{spongy parenchyma}}$	1.5±0.1	1.3±0.1	0.9±0.1	0.7±0.04

Fig. 2. Sections through sun and shade beech leaves.

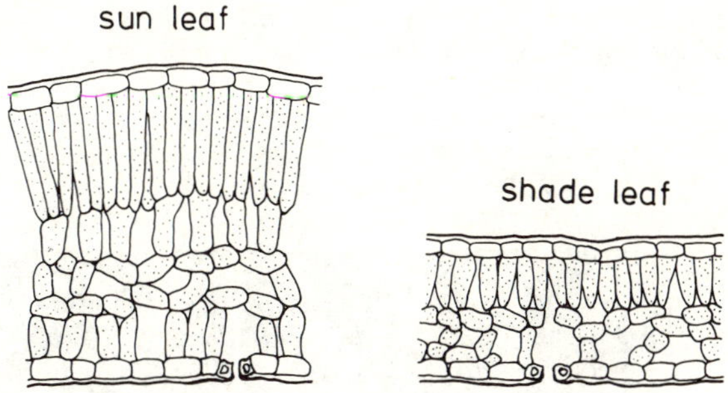

in sun leaves, is always larger than the single cell layer of the palisade parenchyma (Table 1). A very good indicator for the amount of light exposure during the development of a leaf is the ratio of the thickness of the palisade to the spongy parenchyma. This ratio is high in extreme sun leaves (ca. 1.5) and decreases to the values of ca. 0.7 or even lower in extreme shade leaves (Table 1). These differences can be observed in young leaves about two to three weeks after unfolding of the beech leaves, but are more pronounced in fully developed leaves six to seven weeks after unfolding. During a cloudy and rainy spring time the differences in leaf thickness are much less prominent than in sunny periods; under these conditions the typical second layer of palisade cells (funnel cells) is not clearly developed. The sensitivity of leaf thickness to the amount of incident light can be seen from half-shade leaves, which are mainly in the shade but receive some sunshine during a short period of the day. They exhibit only a single row of long palisade cells (ca. 40 μm) and the ratio of palisade to spongy parenchyma cells is 1.2 ± 0.1.

It is of interest that the information to form a sun leaf or shade leaf during unfolding is already present in the bud and depends on the amount of incident light during bud formation in the preceeding autumn (Nordhausen, 1903). This refers to the first leaf formed from a bud. If the branch with a predisposition for sun leaf formation is placed into the shade during leaf unfolding, the plant can

Fig. 3. Scanning electronmicrographs at 77K of freeze-fracture preparations of (A) a sun leaf (ca. 600x) and (B) a shade leaf (ca. 400x) of beech showing the differntial spatial arrangement of cells in the palisade and spongy parenchyma. The upper epidermis of the sun leaf is missing. (Micrographs taken by H.K. Lichtenthaler & K. Oates in 1981 at the University of Lancaster).

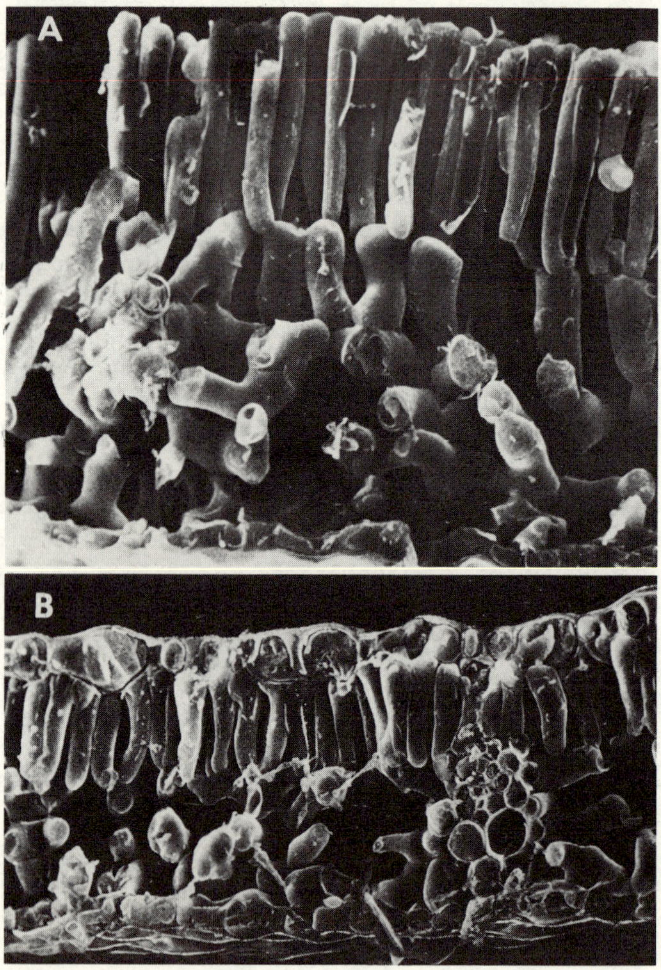

partly adapt to the new light situation and form half-shade or normal shade leaves (Hansen, 1959).

Sun leaves are in a different physiological state than shade leaves as can be seen from their chemical composition. Their metabolism is changed towards the accumulation of a higher proportion of lipids, starch, soluble carbohydrates and cutin whereas shade leaves contain higher relative amounts of protein and soluble amino acids per g of dry matter (Table 2). There are not only differences in the chlorophyll a/b ratio but also in the total chlorophyll content between sun and shade leaves (Table 3 and 4). On a leaf area basis the thicker sun leaves usually contain a greater amount of total chlorophyll, as well as carotenoids and prenylquinones, whereas shade leaves with their relatively low dry weight contain more total

Table 2. Differences in the chemical composition of the dry matter of sun and shade leaves of Fagus sylvatica. After Kausch and Haas (1965).

	sun leaf	shade leaf
compounds (% of dry weight)	17.7	13.3
lipids	1.2	0.8
cutin	4.2	0.8
starch	26.9	37.8
holo-cellulose	14.6	17.3
protein	6.2	5.4
ash		
Elements (mg/100 g dry weight)		
N	2338	2769
K	502	980
Mg	190	252
S	82	96
P	110	117
Ca	888	726

chlorophyll on both a fresh or dry weight basis (Table 3 and 4). In contrast to shade leaves, sun leaves exhibit considerable variations in the chlorophyll content per leaf area unit (Fig. 4a). In the case of extreme sun leaves (sun leaves from the tree top), which possess high chlorophyll a/b ratios (4.5-5.5) the chlorophyll content per leaf unit area is low and may lie beyond that of the average shade leaf (e.g. extreme sun leaf number 2 in Fig 4a).

Table 3. Average chlorophyll content of fully developed sun and shade leaves of Fagus sylvatica. Mean values from 10 leaves harvested in July.

	sun leaf	shade leaf
average leaf area	25 cm	56 cm
fresh weight	0.43 g	0.46 g
dry weight	0.21 g	0.17 g
water content	51%	64%
total chlorophyll content		
mg/leaf	1.2	2.4
mg/100 cm^2	5.6	4.4
mg/g fresh weight	2.8	5.3
mg/g dry weight	5.7	14.4
ratio chlorophyll a/b	3.6±0.2	2.7±0.1
ratio xanthophyll/carotenes	1.6±0.1	2.1±0.2

The level of soluble carbohydrates is a parameter which is very sensitive to changes in the amounts of incident light and reflects the differential photosynthetic rates between sun and shade leaves. One sunny days it is much higher in sun than shade leaves (Table 5). On rainy or cloudy days the amount of soluble carbohydrates is lower in all leaves, and the differences between sun and shade leaves would appear to be less pronounced.

Table 4. Chlorophyll content (a+b) and a/b ratios in sun and shade leaves of different trees. Chlorophyll content is expressed as mg g^{-1} freshweight (fw) and mg g^{-1} dry weight (dw).

	sun leaf			shade leaf		
	fw	dw	a/b	fw	dw	a/b
Aesculus	1.7	4.0	3.5	5.4	19	3.0
Fagus	2.8	5.7	3.6	5.3	14	2.7
Quercus	1.0	2.2	3.8	9.2	23	3.0
Tilia	2.2	6.4	3.1	4.3	16	2.7

Fig. 4. Chlorophyll a/b ratio and light-saturated photosynthetic rates of sun and shade leaves of beech as a function of the chlorophyll content per unit leaf area. m, mean value from nine leaves of each leaf type (after Lichtenthaler et al., 1981).

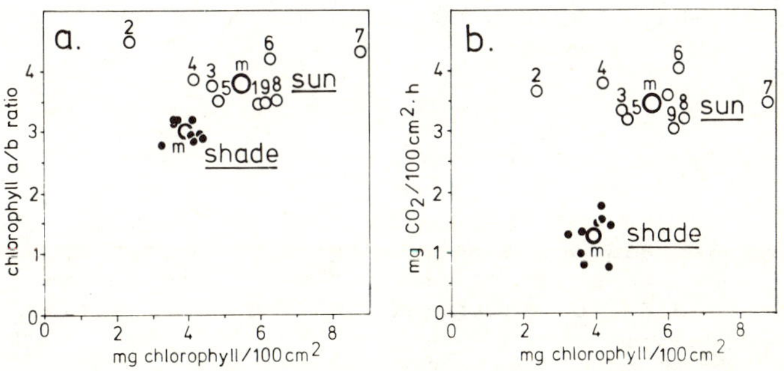

The photosynthetic carbon dioxide fixation rates are higher in sun than in shade leaves on the basis of both leaf area or chlorophyll content (Lichtenthaler et al., 1981) as are the dark respiration rates (Table 6). The average sun leaf also exhibits higher light compensation and light saturation points. Extreme sun leaves with their high chlorophyll a/b ratios possess equal (Fig. 4b) or even higher carbon dioxide fixation rates (Table 6) than the average sun leaf, though their chlorophyll content per leaf area is rather low.

Table 5. Differences in the level of soluble carbohydrates in fully developed sun and shade leaves of four tree species. The values, determined in July, are expressed as μmol glucose equivalents g^{-1} fresh weight (fw) and dry weight (dw) and per cm^2 leaf area (cm^{-2}).

	sun leaf			shade leaf		
	fw	dw	cm^{-2}	fw	dw	cm^{-2}
Fagus						
rainy days[a]	103	227	1.1	50	155	0.4
sunny days[b]	180	370	3.0	120	290	1.0
Aesculus[c]	51	133	1.2	32	112	0.4
Quercus[c]	44	124	1.2	40	110	0.6
Tilia	50	150	1.2	12	48	0.2

[a] minimal values on rainy days; [b] maximum values on days with full sunshine; [c] values taken on a cloudy day.

Table 6. Rates of net photosynthetic carbon dioxide fixation and dark respiration in sun and shade leaves of Fagus sylvatica at 30 klux. Rates are expressed on the basis of unit leaf area and total chlorophyll content and are given as mg $CO_2 . dm^{-2} . h^{-1}$ and mg $CO_2 . mg^{-1}$ chlorophyll. h^{-1} respectively.

	Chl a/b	Rate/unit area		Rate/Chl	
	ratio	phs.	resp.	phs.	resp.
extreme sun leaf	5.0	3.79	0.46	1.30	0.16
sun leaf	3.9	2.96	0.44	0.61	0.09
shade leaf	2.8	0.89	0.19	0.23	0.05
extreme shade leaf	2.7	0.73	0.16	0.18	0.04
ratio sun/shade leaf		3.3	2.3	2.7	1.8
ratio extreme sun to extreme shade leaf		5.2	2.9	7.2	4.0

HIGH-LIGHT AND LOW-LIGHT LEAVES

Under high light (HL) growth conditions plants are characterised by a faster growth and development than in low light (LL) as has been shown for several plants (e.g. Lichtenthaler et al., 1981). HL-radish plants exhibit an increased cotyledon area, a reduced hypocotyl length compared to LL-plants and also an anthocyanin accumulation in the sub-epidermal layers of the hypocotyl which is absent in the LL-seedlings. The appearance and growth of primary and secondary leaves proceeds much faster than in LL-plants (Fig. 5). Leaf and cotyledon growth in LL-plants are limited by the low photosynthesis rates which occur at low light intensities. The fact that the storage material of the

Fig. 5. Appearance of radish seedlings grown for 21 days under high light (60 W m^{-2} PAR, ca. 18 klux) and low light (7 W m^{-2} PAR, ca., 1 klux) on a 14/10 h day/night cycle. Co, cotyledons; 1st and 2nd, primary and secondary leaves.

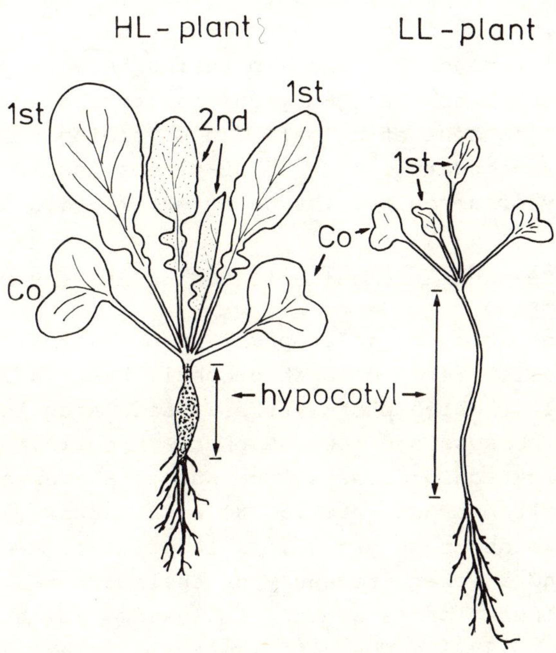

cotyledon is mainly used for the extension growth of the hypocotyl also contributes to reduced cotyledon growth. Due to this reduced growth of the LL-radish leaves and cotyledons, their stomatal density is higher than in the HL-plants.

In many other parameters *e.g.* leaf thickness, second row of palisade parenchyma cells (Fig. 6), high photosynthesis rates or high chlorophyll *a/b* ratio, HL-leaves show the same characteristics as sun leaves, and LL-leaves are comparable to shade leaves (Table 7). This also applies to the leaves of monocotyledonous plants, *e.g.* wheat and maize. In low light their leaves are longer, thinner and also narrower than the shorter and broader HL-leaves. The process of leaf succession, the formation, development and wilting of leaves, proceeds faster under HL than LL growth conditions. This is also reflected in the kinetics of chlorophyll accumulation and breakdown (Fig. 7). In HL-leaves maximum chlorophyll content is reached earlier and the destruction proceeds faster than in LL-leaves. Since during chlorophyll breakdown lower *a/b* ratios arise, the typical differences in the *a/b* ratios between HL- and LL-leaves often show up only during the days of maximum chlorophyll content of a leaf. In physiological terms, the primary or secondary leaves of HL-plants are much older than the primary or secondary leaves of equally old LL-plants.

THE PHOTOSYNTHETIC APPARATUS OF HIGH AND LOW LIGHT LEAVES

The higher chlorophyll *a/b* ratios, the higher photosynthesis rates on a chlorophyll basis and the higher light level at which photosynthesis saturates indicate that HL-leaves possess a different photosynthetic apparatus than LL-leaves. HL-chloroplasts from sun or HL-leaves possess a different ultrastructure (Fig. 8) with narrow grana, only a few thylakoids per granum, a lower degree of thylakoid stacking and a lower frequency of thylakoids as compared to LL-chloroplasts of shade and LL-leaves (Lichtenthaler *et al.*, 1981, 1982). The differences in the ratio of the

Fig. 6. Sections through a HL-leaf and a LL-leaf from high light and low light mustard plants (after Grahl and Wild, 1973).

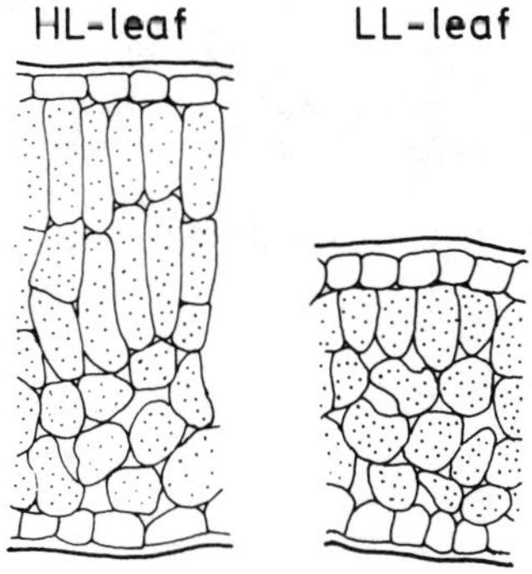

Fig. 7. Differences in the accumulation rate and level of total chlorophyll (a + b) in primary and secondary leaves between high light (HL) and low light (LL) wheat plants (Triticum aestivum L., cv. Kolibri). HL, 90 W m^{-2} (ca. 20 klux); LL, 10 W m^{-2} (ca. 1.5 klux) (after Lichtenthaler et al., 1981).

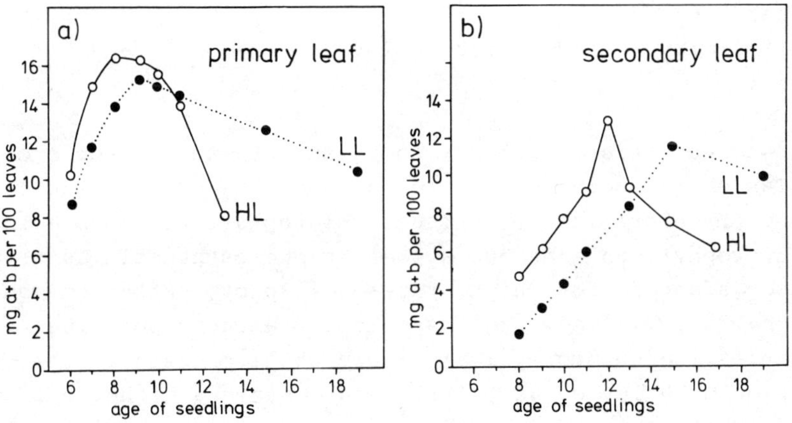

Table 7. Average chlorophyll content (mg per dry weight and per leaf area) and photosynthetic carbon dioxide fixation (per chlorophyll and per leaf area unit) at light saturation (60 klux) of HL- and LL-leaves from several plants. The measurements were performed with fully developed leaves at their maximum chlorophyll content and photosynthetic activity (after Lichtenthaler, 1983).

	Chl a+b		Chl	CO_2 fixation rate	
	mg g^{-1}	mg dm^{-2}	a/b	mg $dm^{-2}h^{-1}$	mg mg^{-1} h^{-1}
Fagus					
sun leaf	7	5	3.9	9.6	1.9
shade leaf	16	4	3.0	3.0	0.8
Raphanus					
HL-cotyledon	5	2.7	3.7	16.2	6.0
LL-cotyledon	17	2.8	2.9	5.1	1.8
Zea mays					
HL-leaves	16	2.6	4.4	23-27	9-11
LL-leaves	27	2.4	3.1	5-15	2-6
Triticum					
HL-primary leaf	12	2.7	3.2	-	-
LL-primary leaf	19	1.8	2.9	-	-

photosynthetic pigments between HL- and LL-chloroplasts (Table 8) indicate a different chemical composition of HL- and LL-chloroplasts. HL(sun)-chloroplasts possess a smaller chlorophyll antenna and a smaller photosynthetic unit, which is adapted for high rates of photosynthetic quantum conversion. They possess lower amounts of the light-harvesting chlorophyll a/b-xanthophyll proteins, the LHCPs (Lichtenthaler et al., 1982) and contain a higher proportion of β-carotene indicative of a higher proportion of the chlorophyll a/ β-carotene proteins CPI and CPIa, which results in the observed higher chlorophyll a/b ratio and in

Fig. 8. Ontogenetic adaptation of the photosynthetic apparatus to high or low quanta fluence rates. P = plastoglobuli. When transferred from one light intensity to the other, chloroplasts will gradually change from HL- to LL-chloroplasts and vice versa within about 2 to 5 days depending on the physiological condition and type of plant.

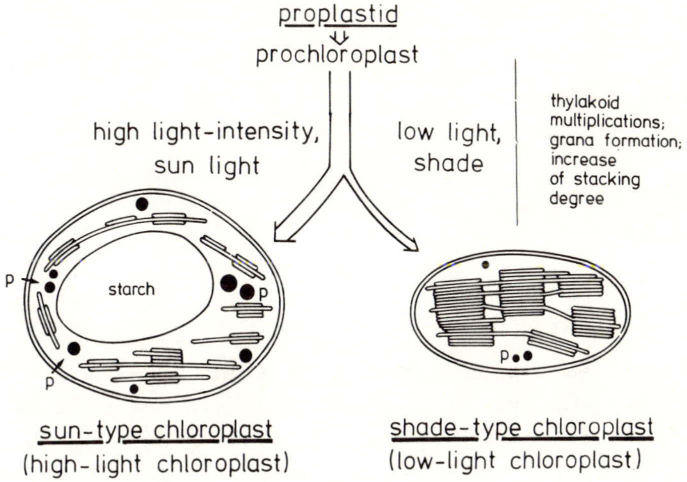

the lower value for the ratio xanthophylls/ β-carotene and total chlorophylls to total carotenoids (Table 8). In contrast, LL(shade)-chloroplasts exhibit a larger photosynthetic unit with relatively greater amounts of LHCPs; they are efficiently adapted for light absorption, but possess a limited capacity for quantum conversion. This adaptive modification of the photosynthetic apparatus is summarised in Fig. 9.

EFFECT OF LIGHT QUALITY

In blue and red light seedlings exhibit many growth responses which are similar to HL(sun)-type and LL(shade)-type growth responses respectively. This mainly refers to the photosynthetic apparatus, the relative content of

Table 8. Pigment ratios (weight ratios) in chloroplasts and isolated pigment protein complexes of radish seedlings (Raphanus sativus L.) grown for 8 days at high (90 W m^{-2}) or low (7 W m^{-2}) light conditions. Pigment proteins were isolated by polyacrylamide gel electrophoresis. (a/b, chlorophyll a/b; c, β-carotene; x, xanthophylls).

	a/b	a/c	x/c	(a+b)/(x+c)
HL-chloroplast	3.8	9.8	1.3	5.4
LL-chloroplast	2.6	17.0	2.2	7.4
Pigment proteins				
CPI + CPIa	9-21	7-10	0.3	6-9
CPa	3-8	4-8	0.4	3-5
LHCP	1.1-1.3	60-180	40-80	3-8

photosynthetic pigments in blue light and red light leaves, the pigment ratios, the photosynthetic activity and chloroplast ultrastructure (Buschmann et al., 1978; Lichtenthaler & Buschmann, 1978). In blue light developing leaves produce sun-type chloroplasts with a higher Hill activity whilst in red light of low or high intensity the formation of shade-type chloroplasts occurs (Lichtenthaler & Buschmann, 1978; Wild, 1979). As in sun and HL-leaves the maximum chlorophyll content is reached earlier in blue light than in red light leaves; this also applies to the chlorophyll breakdown (Lichtenthaler et al., 1980a). A thicker leaf, a higher dry weight or the accumulation of anthocyanins in the cotyledons and hypocotyls can also be seen in blue light plants as compared to red light plants. All differences between blue light and red light leaves are, however, less pronounced than those between sun and shade or HL- and LL-leaves.

CONCLUSION

The largest differences in the morphogenesis of plants, particularly in the anatomy, growth and composition

Fig. 9. Model of the functional organisation of the different chlorophyll-carotenoid-proteins in the thylakoid membrane showing the preferential location of β-carotene and chlorophyll a in the three chlorophyll a-proteins CPa, CPI and CPIa. The chlorophyll a/b-proteins (LHCPs) contain the xanthophylls lutein and neoxanthin; their amounts can be modified by environmental factors as indicated.

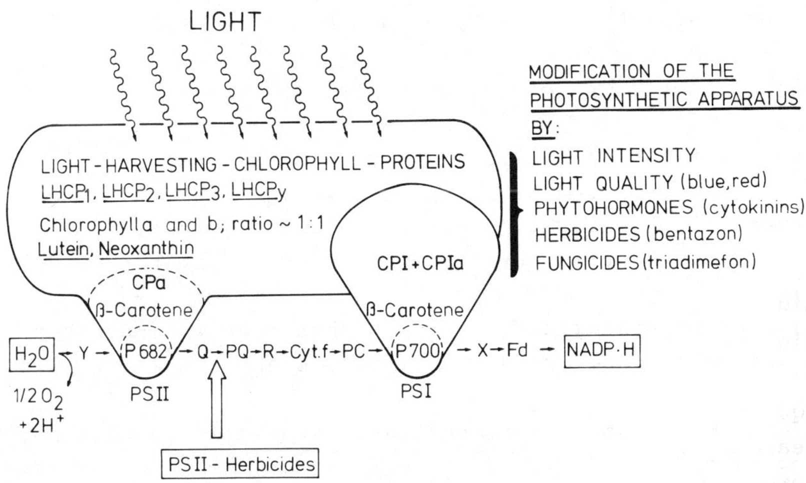

of leaves as well as in chloroplast adaptation, are found between sun and shade leaves and between plants cultivated in high or low light growth conditions. Most of these growth responses, summarised in Table 9, are of physiological advantage for the plant and can be regarded as true adaptations. The adaptation response of leaves and chloroplasts to either high or low quanta fluence rates is the basic central growth response found in plants. The plant has to make use of these two response possibilities when other environmental factors change. Therefore all responses of plant and leaf growth to changes in the temperature, water supply or mineral nutrition as well as to the

Table 9. High-light adaptation characteristics of leaves as found in sun leaves and in leaves of high light plants. The list given is based on the observations in several plants in part by the author and also by other laboratories as indicated below.

thicker leaves and smaller leaf area
higher dry weight and lower water content
longer palisade cells, often 2 rows of palisade cells
higher stomatal density, decreased stomatal diffusive resistance
higher number of chloroplasts per leaf cell

increased carbon dioxide conductivity per unit leaf area
increased photosynthetic carbon dioxide fixation rates
higher light saturation point for photosynthesis
higher level of soluble carbohydrates and starch

higher ubiquinone content and higher respiration rates
higher glycolate oxidase activity, higher level of malate

equal or higher chlorophyll and carotenoid content per unit leaf area
lower chlorophyll and carotenoid content per fresh or dry weight
higher lipid and cutin content per dry weight
lower level of soluble and bound proteins
often lower level of soluble amino acids
increased lignin content, higher level of flavonols
increased level of cytokinins

After Björkman (1975), Boardman (1977), Budde (1923), Butterfass (1979), Fedtke (1979), Kausch & Haas (1965), Lichtenthaler (1971, 1979, 1980, 1981), Lichtenthaler & Buschmann (1978), Lichtenthaler et al. (1981), Wild & Wolf (1980).

application of herbicides, growth regulators, fungicides or other chemicals exhibit great similarity to either a strong light or a weak light growth response of the plant, the leaf and the photosynthetic apparatus. This point is summarised in Fig. 10. Blue light, treatment with cytokinins or the fungicide triadimefon induce a sun-type growth response (Lichtenthaler, 1979; Lichtenthaler et al., 1978), whereas low rates of photosynthesis as induced by photosystem II herbicides result in a shade-type adaptation response (Lichtenthaler, 1979; Lichtenthaler et al., 1980b).

Fig. 10. Scheme showing the adaptation of plant and leaf growth to high and low light quanta fluence rates. At medium light conditions chemicals induce either a sun-type (e.g. cytokinins and the fungicide triadimefon) or a shade-type growth response (e.g. bentazon or other, diuron-type photosynthesis inhibitors).

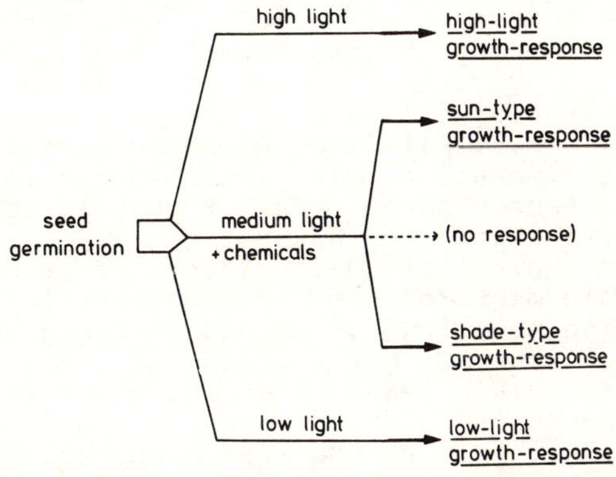

ACKNOWLEDGEMENTS

This work was sponsored by a grant from the Deutsche Forschungsgemeinschaft. I wish to thank Mrs Gabriele Ihrig and Mrs Inge Schürer for assistance.

REFERENCES

Björkman, O. (1975). Environmental and Biological Control of Photosynthesis. In Environmental and Biological Control of Photosynthesis, ed. R. Marcelle, pp. 1-16. the Hague: Dr W. Junk.

Boardman, N.K. (1977). Comparative photosynthesis of sun and shade plants. Ann. Rev. Plant Physiol. 28, 355-377.

Budde, H. (1923). Beiträge zur Anatomie und Physiologie des Blattes auf Grund volumetrischer Messungen. Bot. Arch. 4, 443-487.

Buschmann, C., Meier, D., Kleudgen, H.K. & Lichtenthaler, H.K. (1978). Regulation of chloroplast development by red and blue light. Photochem. Photobiol. 27, 195-198.

Butterfass, T. (1979). Patterns of chloroplast reproduction, pp. 92-97. Vienna: Springer.

Fedtke, C. (1979). Plant physiological adaptations induced by low rates of photosynthesis. Z. Naturforsch. 34c, 932-935.

Grahl, H. & Wild, A. (1973). Lichtindduzierte Veränderungen im Photosynthese Apparat von *Sinapis alba*. Ber. Dtsch. Bot. Ges. 86, 341-349.

Hansen, H.C. (1959). Der Einflussdes Lichtes auf die Bildung von Licht- und Schattenblättern der Buche *Fagus sylvatica*. Physiol. Plant. 12, 545-550.

Kausch, W. & Haas, W. (1965). Chemische Unterschiede zwischen Sonnen- und Schattenblättern der Blutbuche (*Fagus sylvatica* L. cv. Atropunicea). Naturwissenschaften 52, 214-215.

Lichtenthaler, H.K. (1971). Die unterschiedliche Synthese der lipophilen Plastidenchinone in Sonnen- und Schattenblattern von *Fagus silvatica* L. Z. Naturforsch. 26b, 832-842.

Lichtenthaler, H.K. (1979). Effect of biocides on the development of the photosynthetic apparatus of radish seedlings grown under strong and weak light conditions. Z. Naturforsch. 34c, 936-940.

Lichtenthaler, H.K. (1980). Prenylquinones in plant leaves. In Advances in the biogenesis and function of plant lipids, ed. P. Mazliak, pp. 299-310. Amsterdam: Elsevier.

Lichtenthaler, H.K. (1981). Adaptation of leaves and chloroplasts to high quanta fluence rates. In Photosynthesis VI, ed. G. Akoyunoglou, pp. 273-287, Philadelphia: Balaban Internat. Sci. Services.

Lichtenthaler, H.K. (1983). Differences in chlorophyll levels, fluorescence and photosynthetic activity of leaves from high-light and low light seedlings. In Photosynthesis and Plant Productivity, ed. H. Metzner, pp. 194-198, Stuttgart: Wissenschaftl. Verlagsgesellschaft.

Lichtenthaler, H.K. & Buschmann, C. (1978). Control of chloroplast development by red light, blue light and phytohormones. In Chloroplast development, eds. G. Akoyunoglou et al., pp. 801-816. Amsterdam: Elsevier.

Lichtenthaler, H.K., Buschmann, C. & Rahmsdorf, U. (1980a). The importance of blue light for the development of sun-type chloroplasts. In The Blue Light Syndrome, ed. H. Senger, pp. 485-494. Berlin: Springer.

Lichtenthaler, H.K., Burkhard, G., Grumbach, K.H. & Meier, D. (1980b). Physiological effects of photosystem-II herbicides on the development of the photosynthetic apparatus. Photosynthesis Res. 1, 29-43.

Lichtenthaler, H.K., Buschmann, C., Döll, M., Fietz, H.J., Bach, T., Kozel, U., Meier, D. & Rahmsdorf, U. (1981). Photosynthetic activity, chloroplast ultrastructure and leaf characteristics of high-light and low-light plants and of sun and shade leaves. Photosynthesis Res. 2, 115-141.

Lichtenthaler, H.K., Kuhn, G., Prenzel, U., Buschmann, C. & Meier, D. (1982). Adaptation of chloroplast ultrastructure and of chlorophyll-protein levels to high-light and low-light growth conditions. Z. Naturforsch. 37c, 464-475.

Nordhausen, M. (1903). Über Sonnen- und Schattenblatter. Ber. Deutsch. Bot. Ges. 21, 30-45.

Wild, A. (1979). Physiology of photosynthesis in higher plants. The adaptation to light intensity and light quality. Ber. Deutsch. Bot. Ges. 92, 341-364.

Wild, A. & Wolf, G. (1980). The effect of different light intensities on the frequency and size of stomata, the size of cells, the number, size and chlorophyll content of chloroplasts in the mesophyll and the guard cells during the ontogeny of primary leaves of *Sinapis alba*. Z. Pflanzenphysiol. 97, 325-342.

THE MECHANISM OF LIGHT-STIMULATED LEAF CELL EXPANSION

E. Van Volkenburgh, R.E. Cleland, and M.G. Schmidt
Department of Botany, University of Washington,
Seattle, Washington 98195, U.S.A.

INTRODUCTION

One of the most dramatic developmental effects of light on plants is its influence on dicotyledonous leaf expansion. In general, these leaves expand to a very limited extent in the dark, but when they are exposed to light, two different processes occur resulting in 10- to 50-fold increase in leaf surface area (see review by Vince-Prue & Tucker, 1983). Low intensity light acting through the phytochrome system stimulates cell division and some cell enlargement. However, in order to reach full size, leaves must be exposed to bright light which causes the cells of the leaf surface to increase an additional 5- to 10-fold in area (Butler, 1963; Verbelen & de Greef, 1979). We have been primarily interested in the mechanism by which bright light induces this large amount of cell enlargement.

In dicots, cell division generally takes place during early development, while cell enlargement occurs later. There is considerable overlap in both time and localisation of these two processes (Maksymowych, 1973). We have developed a method for temporally separating cell division from enlargement in bean leaves, and with this system we have been able to study physiology of leaf cell enlargement, free from the complexities of cell division. We have used this system to determine how cell enlargement is induced by light, and why growth ultimately ceases as the leaves mature.

THE BEAN LEAF SYSTEM

Primary bean leaves (Phaseolus vulgaris) grown for

10 days after planting in dim red light (RL; 4 μmol m^{-2} s^{-1}) complete normal cell division but do not fully expand. Exposure of these leaves to relatively bright white light (WL; 250 μmol m^{-2} s^{-1}, mixed cool white fluorescent and incandescent) on day 10 stimulates cell enlargement, and causes rapid leaf growth completely free of cell divisions (Van Volkenburgh & Cleland, 1979). Leaves exposed to WL on day 10 reach nearly the same final size as leaves grown up from planting in continuous WL, whereas leaves maintained in RL do not expand significantly after day 10 (Fig. 1). Growth rate is greatest 2 days after exposure to light, then declines until leaves reach full size on day 16.

Fig. 1. Increase in area of intact bean leaves grown in continuous red light (RL), continuous white light (WL), or transferred from red to white light (RL→WL) on day 10 after planting. Data from Van Volkenburgh & Cleland (1979).

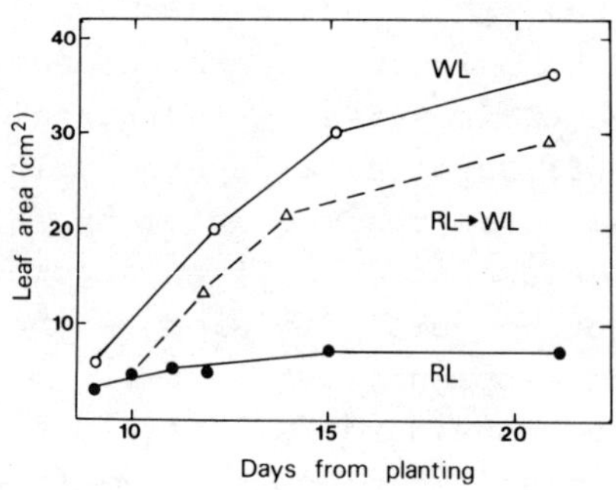

Light also stimulates growth of cells in excised leaf pieces. Discs taken on day 10 from leaves in RL and placed on distilled water in WL expand rapidly for several hours. However, growth can be maintained for up to 24 h by incubating discs on a combination of sucrose (0.01M) and KCl (0.01M), and occurs at a rate comparable to that of intact leaves. Disc expansion after day 10 is also entirely due to cell enlargement (Van Volkenburgh & Cleland, 1979).

Light is known to stimulate many metabolic processes. For example, increased photosynthetic activity may produce energy and precursors for the enzyme and carbohydrate synthesis necessary for growth. Many researchers have correlated the content of metabolites and the rate of synthetic processes with growth rate. However, for any metabolic process to stimulate cell enlargement, it must act by changing one of the physical parameters which determine the rate of cell enlargement.

As has been discussed by Tomos (Chapter 1), plant cell growth may be modelled by the equation derived by Lockhart (1965). Any change in growth rate may then be explained in terms of a change in wall extensibility (ϕ), hydraulic conductivity (Lp), osmotic potential gradient ($\Delta\pi$), and wall yield threshold (Y). While measurements of these cellular parameters in growing leaves has necessarily involved obtaining average values for the complex cell types involved, it has been possible to evaluate changes in each parameter and explain in qualitative terms both the stimulation of cell enlargement by light, and also the cessation of growth as leaves mature.

WALL EXTENSIBILITY

We have used the Instron technique to determine whether wall extensibility (ϕ, here referred to as WEx to specify Instron-measured extensibility) of leaf tissue changes during the induction of leaf expansion by light, or during the cessation of growth upon leaf maturation. With this technique, leaf strips are mechanically extended twice at a constant rate to 20g tension. Comparison of the slopes of the stress/strain curve from the first and second extensions provide a measure of the relative amount of plastic or irreversible extensibility. In theory, this value should indicate the extensibility of the cell walls (ϕ) over the time period immediately before the strips were harvested (Cleland, 1984).

Light causes a rapid stimulation of growth (Fig. 1)

and also increases WEx (Fig. 2). Leaves maintained in RL do not expand, and WEx remains constant (Van Volkenburgh & Cleland, 1980). This suggests that light-stimulated growth may be regulated by mechanisms controlling ϕ. In other tissues, ϕ is thought to be regulated by the acidity of the cell wall, and the acid-growth theory has been developed to explain this regulation (Cleland & Rayle, 1978; Cleland, 1981). This theory suggests that any agent which causes the cell to excrete H into the apoplast stimulates growth rate by increasing ϕ. The acidification of the cell wall, possibly by activating a hydrolytic enzyme, cleaves load-bearing bonds in the wall matrix. This theory has been developed to explain auxin-induced growth of oat coleoptiles, is supported by data from several other growing tissues, and has been tested in growing leaf cells with the following results (Van Volkenburgh & Cleland, 1980).

Fig. 2. Extensibility (WEx), area (○) and growth rate (●) of leaves growing in WL for 6 days. S.e. is indicated, n = 10.

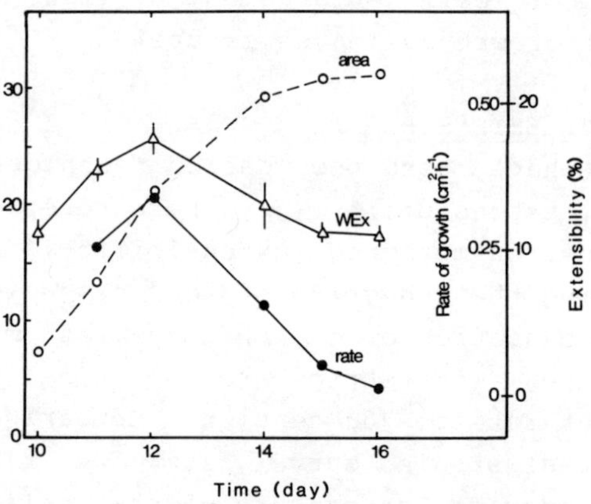

Light causes leaf cells to excrete acid. This is easily measured by placing a flat-tipped combination pH electrode directly onto a lightly-abraded leaf surface, with

0.1 cm^{-3} weak buffer (1 mM KH_2PO_4 - citrate, pH 6) wetting the electrode tip. Leaves maintained in RL have a surface pH of 6.0-6.5. Exposure to WL initially causes a rise in surface pH (probably due to photosynthetic carbon dioxide uptake), followed within 15 min by a steady decrease in pH to a new value of pH 5.0 or less in 2 h (Fig. 3A). The time course of surface acidification corresponds with the initiation of leaf extension (Fig. 3B). This H^+-excretion may also be measured in solution external to leaf pieces (1 mm^2) vacuum-infiltrated in a small volume of 1 mM KCl. Acidification of the external medium is apparent within 60 min of exposure to light, and is inhibited by the protein synthesis inhibitor, cycloheximide (10 $\mu g\ cm^{-3}$), the ATP-ase inhibitor dicyclohexylcarbodiimide (10^{-5}M), or by returning the tissue to RL or darkness.

Fig. 3. Light-stimulated acidification and growth of leaves, and of acid-induced extension of cell walls. (A) pH of abraded leaf surface before and after exposure to white light (WL). (B) Extension of a growing leaf strip after exposure to white light (WL), as measured by position transducer. (C) Extension of isolated cell walls under 10 g tension in response to acidification. Bars indicate 10 min in A and B. Arrow indicates buffer change from pH 7 to pH 6, 5, or 4 in C. Data are from Van Volkenburgh & Cleland (1980).

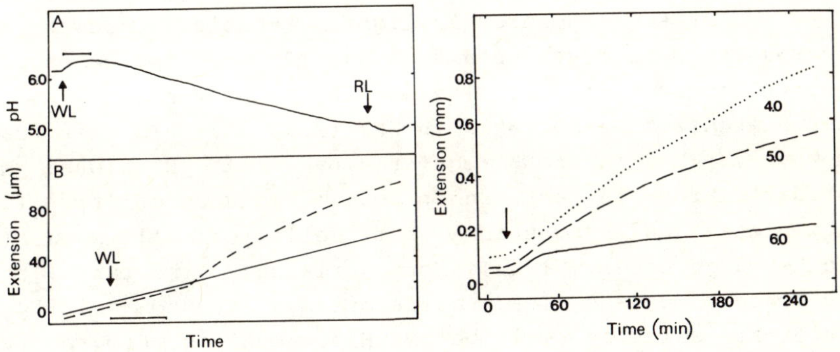

Acidification of isolated cell walls increases their extensibility. Leaf strips, frozen-thawed to remove turgor and disrupt cytoplasmic activity, and placed under 10 g constant tension, extend only slightly when incubated in neutral buffer (0.01 M KH_2PO_4, pH 7). If these strips are acidified by replacing the external buffer with one of a lower pH, extension is induced (Fig. 3C). The rate of acid-induced extension is proportional to pH of the walls. Furthermore, acidification of the cell walls is necessary for growth to occur. Vacuum-infiltration of a neutral buffer (0.01 M KH_2PO_4, pH 7) into the walls of leaf strips growing in solution prevents light-stiulated growth. Infiltration of strips with unbuffered KCl (0.01 M) has no inhibitory effect on growth.

These results support the acid-growth theory, and suggest a mechanism by which light increases leaf growth rate: light stimulates H^+-excretion from the leaf cells, acidifying the cell walls and increasing ϕ. It is not yet known for any tissue, including leaves, how protons are excreted, nor how the cell walls are loosened in response to acidification. Considerable research on the mechanism of auxin-stimulated cell growth has suggested that a plasma-membrane bound ATP-ase pumps H^+ out of the cell (Cleland, 1981), and studies on cell wall chemistry may soon provide an indication of acid-labile linkages in the wall matrix (see papers cited in Taiz, 1984).

After exposure to light, absolute growth rate increases for 2 days, peaks on day 12, then decreases until growth stops on day 16. The decline in growth rate is well correlated with a decrease in WEx (Fig. 2). In this case, however, WEx is not regulated by wall pH alone. Wall acidification is not inhibited in older cells; light stimulates H^+-excretion from cells of all ages (Van Volkenburgh <u>et al</u>., 1984). More strikingly, the fungal toxin fusicoccin causes leaf cells of any age to acidify an external solution well below pH 5, but is progressively less able to stimulate growth as cells age. These results suggest that although acidification of the cell walls may be necessary for growth of young cells, it is not sufficient for growth in older cells.

Fig. 4. Capacity for isolated cell walls to undergo acid-induced extension. Leaf strips (5 X 5 mm) were pre-incubated in neutral buffer (0.01 M KH_2PO_4, pH 7), then extension induced by changing to acidic buffer (0.01 M KH_2PO_4-citrate, pH 4.5). Rate of linear extension was mesured 2-6 h after acidification. S.e. is indicated, n = 25.

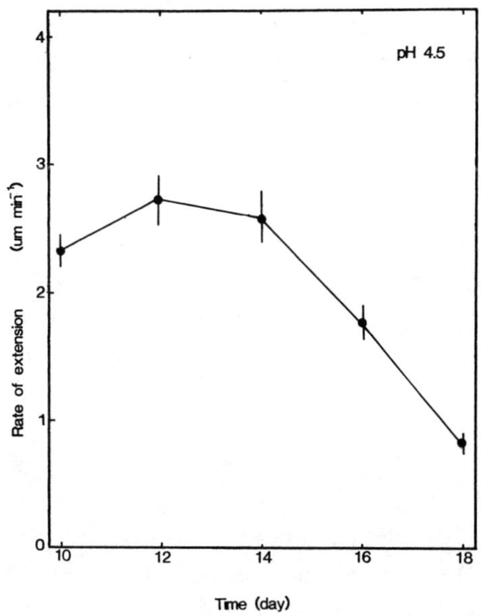

The ability of walls to undergo an increase in WEx in response to acidification may be measured <u>in vitro</u> on frozen-thawed leaf strips as described above (Fig. 3C). Comparison of the rate of strip extension after acidification shows that the capacity for cell walls to extend, when acidified, is gradually lost as leaf cells mature (Fig. 4). These results help to explain the decrease in WEx in older leaf cells, and the limitation of growth in maturing leaves. Similar results have been found in several other tissues when growth is inhibited. Maturation of mung bean hypocotyl cells results in loss of ability to undergo acid-induced growth (Goldberg & Prat, 1981). The reduced ability of Avena coleoptiles to grow in response to fusicoccin after 2-3 h of incubation appears to be due to loss of the capacity of the walls to be loosened by acid (Cleland, 1983). Ethylene inhibition of pea stem growth (Taiz et al., 1983), and the inhibition of leaf growth by abscisic acid (Van Volkenburgh & Davies, 1982) have also been attributed to loss of this capacity.

Thus, ϕ appears to be regulated by two different mechanisms. Acidification of the cell wall is necessary for growth to occur, and initially increases ϕ. This is not however sufficient to maintain growth, and in addition to H excretion, the capacity for the cell wall to be loosened by acid must be maintained. Loss of this capacity may be due to loss of polylamelate structure (Roland et al., 1982), increased isodityrosine crosslinks (Cooper & Varner, 1983; Fry, 1983), or an increase in deferulic acid links (Smith & O'Brien, 1979). It is also possible that the wall-loosening enzyme is not re-synthesised in older cells.

TURGOR PRESSURE AND WALL YIELD THRESHOLD

Growth rate is dependent on the magnitude of turgor pressure (P), which is itself dependent on many variables (Cosgrove, 1981; see also Chapter 1) including cell osmotic potential (π) and membrane hydraulic conductivity (Lp). The water potential gradient ($\Delta\psi$) generated by π determines the maximum driving force for growth (P) available by solutes pulling water into the cell. Concentration of solutes in the cell would potentially increase P, dilution of solutes would reduce P. On the other hand, if Lp is limiting water uptake into growing cells, the resistance to water flow will not allow P to develop to a maximal value. In this case, an increase in Lp would allow water to move into the cell more freely, and P would increase. In this way, changes in either π or Lp may affect P, and therefore modify the growth rate.

Growth induced by WL on day 10 is accompanied by a

Fig. 5. Osmotic concentration (OC) of intact leaves during leaf expansion. Sap was expressed from frozen-thawed leaves and measured in a vapour-pressure osmometer, S.e. is indicated, n = 5.

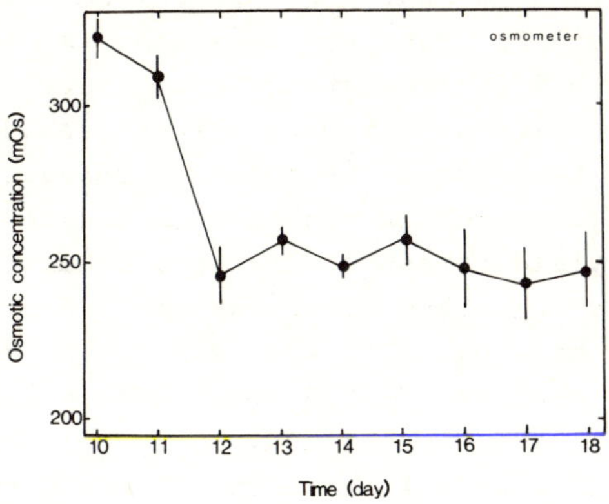

rapid decrease in osmotic concentration (OC) (Fig. 5), which is a result of dilution of solutes by incoming water. (OC is proportional to π: 40 mOsm = 0.1 MPa.) After day 12, growth rate steadily declines, but OC remains constant (Van Volkenburgh et al., 1984). The rapid dilution of solutes which occurs before day 12 contributes to the reduction in P, measured psychrometrically, which is observed during 24 h after stimulation of growth by WL (Van Volkenburgh & Cleland, 1981; Davies & Van Volkenburgh, 1983). The decline in growth rate after day 12 is not associated with a change in OC, and P increases slightly (Van Volkenburgh et al., 1984). Thus, the changes in π which occur are negtively correlated with growth rate, and do not cause P to change in such a way to achieve the observed growth rates.

Lp can be estimated by measuring cell growth rate and the water potential gradient ($\Delta\psi$) from the outside to inside of the growing cells. If $\Delta\psi$ is considered to be the difference between cell water potential (ψ_c) and xylem water potential (ψ_o), and ψ_o is maintained constant, the gradient may be approximated by ψ_c. Measurements of ψ_c and cell growth rate before and after WL-exposure on day 10 indicated that a small but insignificant increase in Lp may occur in response to light Van Volkenburgh & Cleland, 1981. Such a change could in principle allow for an increase in P, but measurements show that P declines (see above). Similar measurements of Lp have not been made as leaves mature. While this method for measuring Lp is very indirect, and should be confirmed with a more direct method, such as the pressure probe technique, the observed changes in P indicate that Lp is not limiting growth in this tissue.

Wall yield threshold (Y) is another parameter which might possibly regulate leaf growth rate. P is only effective in driving wall extension to the extent that it exceeds Y. For example, if 0.4 MPa are necessary to initiate wall extension, and P = 0.5 MPa, only 0.1 MPa exists above Y. However, if Y were lowered to 0.2 MPa with P remaining constant at 0.5 MPa, the turgor effective in driving growth (P-Y) will have tripled, and growth rate would increase.

Indirect measurements of Y have been made by floating leaf strips on solutions of the impermeant solute polyethylene glycol (MW 4000). It has been assumed that reducing the external water potential caused an equivalent drop in cell turgor, and in this way it was possible to determine how much pressure was necessary for growth to occur, or Y. Using this technique, Y was found to be 0.25 MPa in growing bean leaves, and Y did not change when light stimulated growth (Van Volkenburgh & Cleland, 1981), or when age inhibited growth (unpublished results). However, this technique is subject to a number of errors. Firstly, no absolute measure of P is possible. Secondly, the time required for growth to occur and to be measured allows many changes to occur in the tissue, possibly including changes in OC and even shifts in WEx and Y.

Recently, a more direct method of measuring Y has been proposed (Cosgrove _et al._, 1984) in which turgor is allowed to drop to Y after growing cells are excised (_i.e._ removed from their water supply) and wall relaxation continues in the absence of water entry into the cells. P may be measured directly with a pressure probe, or indirectly by a psychrometer. With this method, Y has been found to be approximately 0.3 MPa for leaves growing slowly in RL, and decreases slightly (less than 0.1 MPa) when growth is stimulated by WL. However, the turgor effectively driving growth, P - Y, remains constant because P also drops by about 0.1 MPa when leaves are exposed to WL (Van Volkenburgh & Cleland, 1984). These results indicate that Y is not limiting growth in bean leaves, and also that the change in Y does not cause growth rate to increase, because it is counterbalanced by the change in P.

CONCLUSIONS

The bean leaf system has provided an opportunity to investigate the physical constraints on cell enlargement free from complicating physiology of cell division. Light, which is necessary for the development of dicotyledonous leaves, stimulates rapid cell growth by activating a wall-loosening mechanism, and increasing ϕ. Lp and Y do not change sufficiently to affect growth rate. Similarly, the decrease in growth rate which occurs as leaves mature is associated with loss of ϕ. The cellular mechanisms which regulate ϕ are influenced by light and leaf age, and are therefore important processes controlling leaf growth.

These results indicate that the acid-growth theory applies to leaf cell expansion as well as to hormone-induced growth systems. Light causes H^+-excretion by cells into the wall space. Lowering of wall pH from 6.5 to below 5.0 stimulates wall-loosening and growth, and is necessary for growth to occur. The increased H^+-excretion caused by light sufficiently accounts for the wall-loosening associated with light-stimulated growth. However, H^+ excretion is not sufficient to maintain growth in ageing cells, and maintenance of the capacity for walls to be loosened by acid is also necessary. Older leaf cells are fully capable of excreting H^+, yet they do not grow in response to acidification. In older cells, ϕ is reduced because cell walls lose the capacity to extend in acid. These results suggest that there are at least two steps controlling ϕ; H^+-excretion and maintenance of wall-loosening capacity.

While the regulatory role of ϕ has been demonstrated well in the bean leaf model system, it is quite possible that changes in turgor can still be a major factor in the control of leaf growth under other circumstances. For example, leaf discs floating on water expand in the light for only a few hours in the absence of external solutes, presumably due to a decline in P during growth. If solutes are present so as to permit some maintenance of sufficient π, and thus P, growth is maintained for a much longer period (Van Volkenburgh & Cleland, 1979). In this case, turgor maintenance is necessary for growth to continue. A second example may be taken from the diurnal growth patterns of leaves. Because light-stimulated H^+-excretion is necessary for growth, one might predict that leaves should grow faster during the day than at night. This prediction is confirmed in a number of studies (Christ, 1978; Davies & Van Volkenburgh, 1983; Matthews et al., 1984; Radin & Boyer, 1982), but in other cases growth is faster at night (Boyer, 1968; Radin & Boyer, 1982). It appears that in the latter cases, transpiration combined with high root resistance results in reduced leaf P which inhibits growth during the day. At night, rapid enlargement may occur in response to restored P until the light-induced acidification of the apoplast disappears (a process that may take hours). Thus, P may become limiting to growth under field conditions, and regulation of π through osmotic adjustment may play an important role in maximising growth.

The importance of the physiological control of cell wall properties has been demonstrated for bean leaf cell growth. Several questions arise from the results presented above. Perhaps the most obvious is which photoreceptor is responsible for converting light energy into H^+-excretion, and where does it reside in the leaf. Because it is difficult to obtain high intensity light at narrow band wavelengths, an action spectrum for this response has not yet been determined, and white light has been used.

Experiments are in progress to address this problem.

A second problem is that of applying a cellular model to tissue of mixed cell types. Although it is likely that the epidermal cells limit leaf expansion, it is quite possible that the mesophyll cells are responsible for acidification of the apoplast. Furthermore, osmotic concentration may differ from cell to cell, such that turgor of epidermal cells is not that calculated from bulk tissue measurements. Other questions centre on the mechanism of acidification and the nature of the wall-loosening process itself. Research into these areas is progressing in other growth systems, e.g. auxin-induced coleoptile growth, and questions now can also be asked for growing leaf cells.

Finally, it should be pointed out that the physiology of dicot leaf growth may be quite different from that of monocots. Two major differences are that cell division and enlargement are spacially separated in monocot leaves, and that these processes are not light-stimulated (Vince-Prue & Tucker, 1983). However, the mechanism of cell enlargement may be similar between the two leaf types. Evidence suggests that maize leaf cells have an acid growth mechanism, and that ϕ, rather than P, regulates growth rate (Van Volkenburgh & Boyer, 1984).

REFERENCES

Boyer, J.S. (1968). Relationship of water potential to growth of leaves. Plant Physiol., 43, 1056-62.

Butler, R.D. (1963). The effect of light intensity on stem and leaf growth in broad bean seedlings. J. Exp. Bot., 14, 142-52.

Christ, R.A. (1978). The elongation rate of leaves. J. Exp. Bot., 29, 603-10.

Cleland, R.E. (1981). Wall extensibility: hormones and wall extension. In Plant carbohydrates II: Extracellular carbohydrates, eds. W. Tanner and F.A. Loewus, Encyclopedia of Plant Physiology, New Series. Vol. 13B, pp. 255-76. New York: Springer-Verlag.

Cleland, R.E. (1983). The capacity for acid-induced wall loosening as a factor in the control of *Avena* coleoptile cell elongation. J. Exp. Bot., 34, 676-80.

Cleland, R.E. (1984). The Instron technique as a measure of

the immediate past wall extensibility. Planta 160, 514-520.
Cleland, R.E. & Rayle, D.L. (1978). Auxin H^+-excretion and cell elongation. Bot. Mag. Spec. Issue, 1, 125-39.
Cooper, J.B. & Varner, J.E. (1983). Insolubilisation of hydroxyproline-rich cell wall glycoprotein in aerated carrot root slices. Biochem. Biophys. Res. Comm., 112, 161-67.
Cosgrove, D.J. (1981). Analysis of the dynamic and steady-state responses of growth rate and turgor pressure to changes in cell parameters. Plant Physiol., 68, 1439-46.
Cosgrove, D.J., Van Volkenburgh, E. & Cleland, R.E. (1984). Stress relation of cell walls and yield threshold for growth: demonstration and measurement by micropressure probe and psychrometer techniques. Planta 162, 46-54.
Davies, W.J. & Van Volkenburgh, E. (1983). The influence of water deficit on the factors controlling the daily pattern of growth of Phaseolus trifoliates. J. Exp. Bot., 34, 987-99.
Fry, S.C. (1983). Feruloyated pectins from the primary cell wall: their structure and possible functions. Planta, 157, 111-23.
Goldberg, R. & Prat, R. (1981). Development of the responses to growth regulators during the maturation of mung bean hypocotyl cells. Physiol. Veg., 19, 523-32.
Lockhart, J.A. (1965). An analysis of irreversible plant cell elongation. J. Theor. Biol., 8, 264-75.
Maksymowych, R. (1973). Analysis of leaf development. Cambridge: Cambridge University Press.
Matthews, M.A., Van Volkenburgh, E. & Boyer, J.S. (1984). Acclimation of leaf growth to low water potentials in sunflower. Plant Cell Environ., 7, 199-206.
Radin, J.W. & Boyer, J.S. (1982). Control of leaf expansion by nitrogen nutrition in sunflower plants. Plant Physiol., 69, 771-75.
Roland, R.C., Reis, D., Mosiniak, M. & Vian, B. (1982). Cell wall texture along the growth gradient assembly and dissipative process. J. Cell Sci., 56, 303-18.
Smith, M.M. & O'Brien, T.P. (1979). Distribution of autofluorescence and esterase and peroxidase activities in the epidermis of wheat roots. Aust. J. Plant Physiol., 6, 201-19.
Taiz, L. (1984). PLant cell expansion: regulation of cell wall mechanical properties. Ann. Rev. Plant Physiol., 35, 585-657.
Taiz, L., Rayle, D.L. & Eisinger, W. (1983). Ethylene-induced lateral expansion in etiolated pea stems. Plant Physiol., 73, 413-17.
Van Volkenburgh, E. & Cleland, R.E. (1979). Separation of cell enlargement and division in bean leaves.

Planta, 146, 245-47.

Van Volkenburgh, E. & Cleland, R.E. (1980). Proton excretion and cell expansion in bean leaves. Planta, 148, 273-78.

Van Volkenburgh, E. & Cleland, R.E. (1981). Control of light-induced bean leaf leaf expansion: role of osmotic potential, wall yield stress, and hydraulic conductivity. Planta, 153, 572-77.

Van Volkenburgh, E. & Boyer, J.S. (1984). Inhibitory effects of water deficit on maize leaf elongation. Plant Physiol., 77, 190-194.

Van Volkenburgh, E. & Cleland, R.E. (1984). Calculation of the wall yield threshold of growing bean leaves from isopiestic psychrometer measurements. Plant Physiol., 75 Suppl., Report 56 160.

Van Volkenburgh, E. & Davies, W.J. (1983). Inhibition of light-stimulated leaf expansion by abscisic acid. J. Exp. Bot., 34, 835-45.

Van Volkenburgh, E., Schmidt, M.G., Cleland, R.E. (1984). Limitation of growth in maturing bean leaves. Planta (in press).

Verbelen, J.P. & de Greef, J.A. (1979). Leaf development of Phaseolus vulgaris L. in light and in darkness. Amer. J. Bot., 66, 970-76.

Vince-Prue, D., Tucker, D.J. (1983). Photomorphogenesis in leaves. In The growth and functioning of leaves, eds. J.E. Dale and F.L. Milthorpe. pp. 233-70. Cambridge: Cambridge University Press.

LEAF GROWTH AND WATER DEFICITS: BIOPHYSICAL EFFECTS

Theodore C. Hsiao, Wendy K. Silk, and Jiahai Jing
Department of Land, Air and Water Resources,
University of California, Davis,
California 95616, U.S.A.

INTRODUCTION

Water stress is ubiquitous to life of terrestrial plants, even under conditions of frequent rainfall (Coster, 1927). The most luxuriant vegetation is found in the wet tropics whereas great restriction of foliage growth occurs under arid conditions. Although gardeners and farmers appreciate the importance of water in maintaining good growth, the critical role played by water and the extreme sensitivity of leaf expansion to water stress are usually not recognised. More and more data are now accumulating to indicate that restriction of growth of foliage canopy is the first sign of water stress, often before a reduction in leaf water potential (ψ) can be detected. Measurable leaf-to-leaf and plant-to-plant variations in ψ exist in the field, making it difficult to detect the small differences in at the onset of water stress. Yet a difference in only tens of KPa in ψ can make a substantial difference in leaf expansion rates, as will be elaborated in later sections.

Sustained expansive growth is the result of interactions among numerous processes in the plant, both biophysical and biochemical. As discussed in the other chapters of this book, these processes are affected by factors other than water. Here we will focus our attention on water status and emphasise the more biophysical aspects. Biochemical changes elicited by water stress are addressed by Lawlor and Leach (Chapter 13). This chapter first briefly discusses the current basic biophysical concept of expansive growth, which is more extensively examined in Chapter 1. This is followed by a section on the spatial variation of growth parameters. Only when the spatial variation is appreciated and quantified can we apply the biophysical principles rigorously to the analysis of growth and the effects of water stress. The next section deals with effects of and responses to water stress. Some emphasis is placed on how the time frame of stress alters the parameters underlying growth responses. Also considered is the interdependence of the various parameters once a perturbation initiates a change in one parameter. In addition, the manifestation and implications of the high sensitivity of expansive growth to water stress in the field are discussed. The chapter concludes with comments on some outstanding questions needing further research.

BIOPHYSICAL FRAMEWORK FOR THE ROLE OF WATER

Water makes up 85% or more of the volume of meristematic and expanding zone of the leaf in many species. It follows that water uptake by the growing region is an integral part of the processes sustaining volume expansion and growth. For the simplest case of a single cell, water uptake rate is a function of the ψ differential between the cell exterior and interior; hydraulic conductivities of the cell wall, membranes, the cytoplasm, and the vacuole; and geometry of the cell as it affects path length and cross sectional area for water transport. Hydraulic conductivities are almost never known for parts of a cell. Water uptake inflates the protoplasm confined within the cell wall. If the wall does not give or yield by stretching and relaxation, turgor pressure (P), and hence ψ, would rise in the cell with water uptake to the point where ψ inside equals ψ outside and no further uptake would take place. What distinguishes a growing cell from a mature cell, however, is the ability of its wall to yield and stretch continuously and irreversibly under force normally provided by adequate turgor pressure. This permits the sustained water uptake and cell expansion as long as the wall does not lose its ability to yield and solutes are taken up or generated so cellular osmotica are not diluted sufficiently to erase the differential in ψ. A momentary yielding of cell wall can be purely physical. For sustained stretching, however, wall metabolism has obviously a key role, and continuous deposition of new wall material is necessary. In most cases, acidification of the wall promotes its yielding. Wall metabolism has been reviewed recently by Labavitch (1981).

A biophysical model for water-driven cell enlargement was proposed by Lockhart (1965) and is described in detail in Chapter 1. The model consists of two rather simple equations. One focuses on the wall yielding characteristics, and the other on the water conducting characteristics of the expanding cell. The relative rate of volume (V) enlargement of the cell may be expressed by a combination of the two equations (see Chapter 1). A modified version of the combined equation is

$$\frac{1}{V}\frac{DV}{dt} = \frac{\phi L}{\phi + L}\left(\psi_e - \pi_i - Y\right) \tag{1a}$$

$$= \frac{\phi L}{\phi + L}\left(\Delta\psi + P - Y\right) \tag{1b}$$

where ϕ is volumetric wall extensibility, L is volumetric hydraulic conductance of the cell, and Y is the yield threshold pressure, below which the force on the wall is insufficient to cause wall relaxation. The other parameters in the equation are water potential exterior to the cell, ψ_e, difference in water potential between the cell exterior and interior, $\Delta\psi$, osmotic potential of the interior, π_i, and turgor pressure in the cell, P.

There is some agreement that for single cells and slowly growing tissues, L is so large in relation to ϕ that only ϕ needs to be considered and $\Delta\psi$ is nearly zero (Ray _et al._, 1972; Tyree & Jarvis, 1982; Cosgrove & Cleland, 1983). Equation 1b then reduces to the first equation of Lockhart

$$\frac{1}{V}\frac{dV}{dt} = \phi\left(P - Y\right) \tag{2}$$

Tomos discusses some qualifications which must be made for use of the Lockhart model and points out that it is probably inadequate to explain details of long-term and transient behaviour. It must be kept in mind, therefore, that the model is a first order approximation of a very complex and dynamic system. Thus the physiology of the effect of water stress on growth can be examined, also in first order terms, via analysis of the parameters of the Lockhart model. This is done in a section to follow. Before discussing effects of water stress on leaf growth, however, we must digress to consider the problem of temporal and spatial variation.

CONCEPTUAL FRAMEWORK FOR ANALYSIS OF TEMPORAL AND SPATIAL VARIATION IN THE LEAF

Chapter 1 and the preceding section introduced the currently accepted model for growth sustaining water relations of a plant cell. To generalise this model for the continuum of cells which form the leaf, we must consider some new aspects of growth and water relations. Each leaf cell experiences displacement (due to growth of other cells) as well as its own expansion. Analysis of change in a cell which moves during its period of expansion is a subtle problem which requires that we consider both growth analysis in the leaf and spatial variation in any water related property of interest (Silk & Erickson, 1979).

Temporal Changes in the Leaf as a Whole. Temporal change in the whole leaf is simpler to analyse, so we will begin here with a brief review of the use of a developmental index to study growth affected by water stress. Then we will discuss spatial-temporal interrelationships within a growing leaf.

Properties associated with the whole leaf often increase with development age. Thus, leaf size, amount of leaf chlorophyll and rate of O_2 evolution per leaf increase as the leaf grows. A shoot bearing leaves of different plastochron ages provides material to analyse the developmental pattern of a variable of interest. The variable can be assayed for each leaf and plotted against leaf plastochron age to uncover the developmental pattern. The leaf plastochron index, a continuous developmental index based on leaf serial number (Erickson & Michelini, 1957), provides a useful, quantitative method to indicate the developmental age of a leaf (see Chapter 3).

In addition to revealing many developmental trends, use of the leaf plastochron index can permit assessment of the effect of water stress. For example, in a study of well watered and poorly watered canteloupe vines (Silk, 1980), plots of leaf length versus time reveal that growth is slowed by water deficiency. Plots of leaf length versus leaf plastochron index reveal that for two irrigation treatments, leaf length is synchronised with developmental stage for the first half of the expansion (Fig. 1). Only when the leaf is perhaps 60% expanded does growth of the water deficient leaves slow relative to plastochron age.

The leaf plastochron index has provided a developmental time scale for assessing trends in whole leaf physiology and morphology. It is less useful in analysing spatial variation within the leaf. For example, it does not provide insight into leaf cell age when this varies within a leaf. Below we will consider spatial and temporal variation first in one dimension, in the monocot leaf. Then we will consider the more general problem of two dimensional variation, in a typical broad leaf.

<u>Growth Rates, Velocities and Trajectories - The Monocotyledonous Leaf</u> To describe the distribution of growth in an expanding leaf, it is desirable to have a formulation which is a "point function", that is, an expression which can be evaluated at points, or infinitesimally small increments, on the leaf surface. Such an expression is the relative elemental growth rate (Erickson & Goodard, 1951), also known to fluid dynamicists as the strain rate (Silk & Erickson, 1979) and here called, for simplicity, the growth rate. To understand the distribution of cell expansions which produces the growth velocity of the leaf tip, one

Fig. 1. Temporal variation in leaf length for irrigated and water stressed cantaloupe vines. Upper diagram: length of the fourth leaf of well watered (asterisks) and poorly watered (diamonds) plants plotted against time. Data from 20 plants are shown to illustrate the scatter which results when a developmental variable is observed as a function of time. Lower diagram: leaf length plotted against plastochron index. Use of a developmental index reduces scatter and reveals that stressed leaves slowed their growth, relative to branch development, near the end of the period of leaf expansion. (From Silk, 1980.)

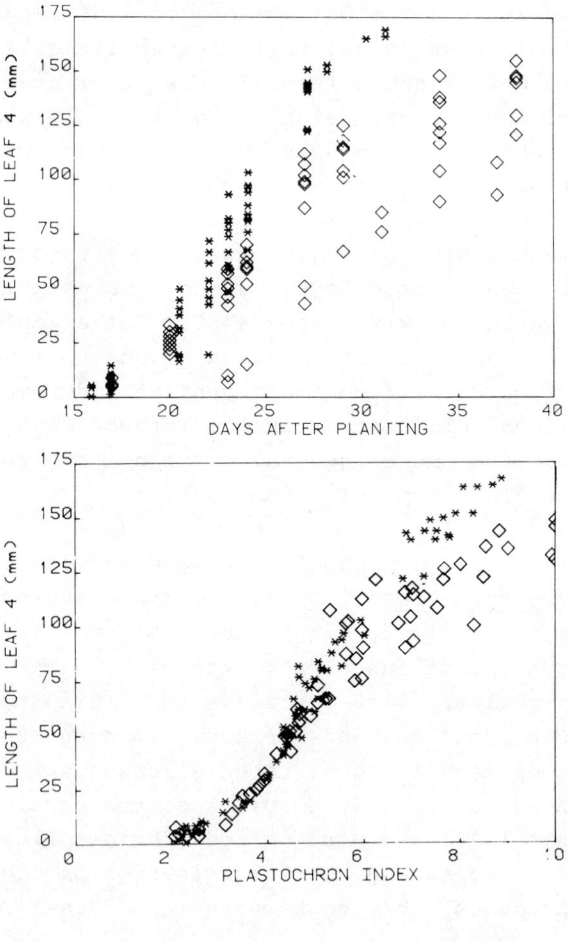

considers the behaviour of a tiny segment with apical end A and basal end B. If A is moving faster than B, then the segment is expanding, and a quantitative description of the segment expansion rate is given by the difference in velocities, v(a)-v(b). This measure of expansion is awkward because it depends on the amount of tissue in the segment. To circumvent the dependence on segment length, one divides by the length of the segment to obtain a relative growth rate (in units of length per unit length per hour) and then performs the calculus trick of shrinking the segment to a point. Mathematically, the resulting expression is known as the velocity gradient. An important implication is that the growth rate distribution can be obtained by differentiating an empirically determined curve of growth displacement velocity versus distance from the meristem.

For the partially expanded fifth leaf of _Zea mays_ growing in a controlled environment chamber, we calculated the growth rate to have a maximum of 0.07 h^{-1} at a spot 2 cm from the stem node. Expansion ceases (the growth zone ends) at a location approximately 6 cm from the node (Fig. 2, upper diagram). The plot of growth rate versus position reveals the spatial distribution of elongation growth in the maize leaf and provides the basic empirical information which the physiologist must attempt to explain.

We have seen that evaluating relative elemental growth rates involves measuring the distribution of growth velocity. A related idea is that the position of a cellular particle can be located by integrating its growth velocity over time. This leads to the notion of a growth trajectory, which can be visualised as a plot of cell location as a function of time or cell age. The growth trajectory of a leaf cellular particle found initially adjacent to the basal meristem is shown in the lower diagram of Fig. 2 which reveals that the particle spends about 65 h reaching the location 2 cm from the node and then only 23 additional hours to be displaced through the remaining 4 cm of the growth zone.

If growth is steady (invariant with time; see Silk,

Fig. 2. Growth analysis for the fifth leaf of maize. Upper diagram: growth velocity and growth rate distributions. Growth velocity (solid line) represents for the continuum of leaf particles the instantaneous rates of displacement from the stem node. Numerical differentiation of the plot of growth velocity versus position gives the growth rate (strain rate, dashed line) which is a basic description of the magnitude and location of growth in the leaf. The leaf was 30 cm long at the time of the analysis. Lower diagram: growth of a leaf tissue element followed in time. The growth trajectory (solid line) of a single leaf particle is a plot of particle position versus time. Knowing the growth trajectory, one can infer the temporal distribution of a variable from spatial patterns. Here the growth rate of the particle (dashed line) is shown to increase slowly for 60 h and to decline abruptly during 12 h at the end of the growth period.

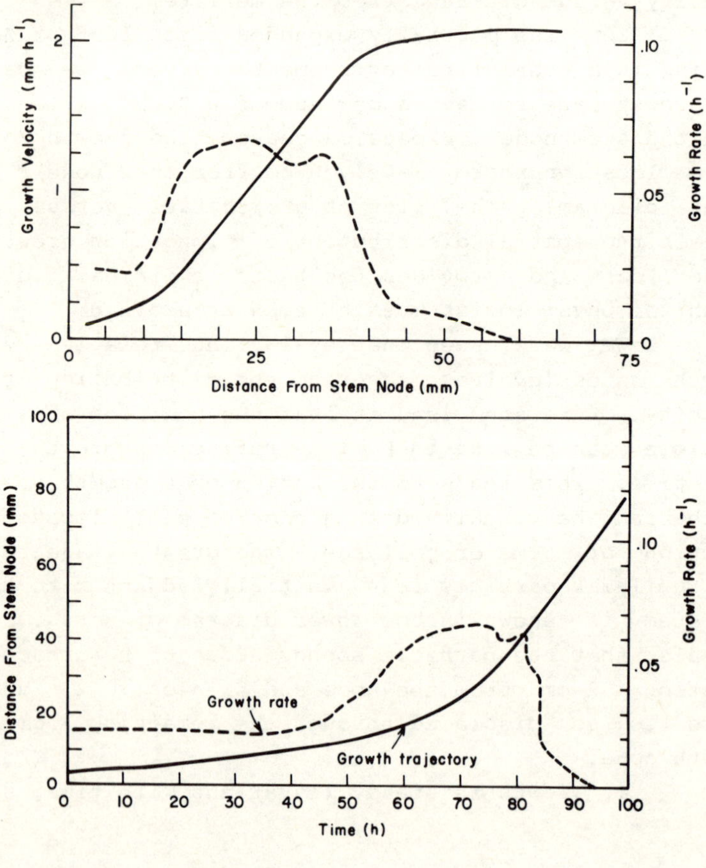

1984), one can combine knowledge of the growth trajectory with data on the spatial distribution of a physiological variable to determine the time course of change in the variable associated with the particular tissue element. As an example, we show the growth rate of the leaf cellular element as a function of time (Fig. 2, lower diagram). This approach is extremely general. Leech and co-workers (Boffey *et al.*, 1980) used similar reasoning in establishing chlorophyll synthesis rates and *a/b* ratios as a function of cell age. In a later section we use the growth trajectory to infer, from the spatial distribution of water potential, the temporal pattern of water potential of the cellular elements.

The Dicotyledonous Leaf. The spatial variation and developmental history of a broad leaf are challenging to analyse. Even if the leaf changes shape and physiological properties in a perfectly predictable way, and if increase in thickness in negligible, expansion is, at the simplest, a two-dimensional process. Growth may be visualised as the expansion of a tiny circle painted on the surface. The relative elemental growth rate, which in one dimension is the relative rate of increase of a tiny line segment, must be visualised in two dimensions as the rate at which a circle deforms into a slightly larger ellipse. Mathematically, the relevant expression is the divergence of velocity which is equal to the relative rate of increase in area of a tiny circle (Richards & Kavanagh, 1943). The relative rate of area increase is given by the sum of the lengths of the axes of the ellipse; the direction and magnitude of maximum expansion are given by the direction and length, respectively, of the major axis of the ellipse. (The minor axis of the ellipse gives the direction and magnitude of the minimum expansion rate which, according to engineering mechanics, is necessarily at right angles to the direction of maximum expansion). When a field of such ellipses is calculated from data on the growth velocities of points on the leaf surface, a picture is obtained of the spatial distribution of the growth rate.

The methodology for determining the spatial distribution of the growth rate in two dimensions is formidable, requiring computer assisted analysis of extensively marked time lapse photographic records. The growth trajectories themselves have x- as well as y-components. As yet, the growth of only a few leaves has been analysed in these quantitative terms. We have pictures of the growth rate distribution of Xanthium at a half-expanded stage, Cucurbita at several developmental stages and Vitis at a partially expanded stage. Putting together these data with earlier qualitative descriptions of leaf growth, we see that for broad-leaved species it is often found that growth rate is greater in basal or interior regions and slower in apical portions. As the leaf develops, the tip and marginal meristem are often the first to mature (see Chapter 3 and Silk, 1983).

The Importance of Understanding Spatial Variation in the Study of Water Stress. To characterise the effect of water deficits on leaf growth, we should know the spatial distribution of the growth response. This could be most definitely accomplished by comparing the growth rate distribution in control and water-stressed plants. Knowing the magnitudes and locations of the effect, one could then correlate changes in cellular π and P with other growth related variables to establish a chain of casuality. Some problems arise in generalising the single cell model to the continuum of cells which comprise the leaf (discussed below). Nevertheless, knowledge of the spatial distribution of the response facilitates comparing local effects to putative local causes. Furthermore, to understand the effect of the water deficit on development, growth trajectories in control and stressed plants should be obtained. Then we can begin to assemble the time course for developmental effects of water deficits.

GROWTH SUSTAINING WATER POTENTIALS IN AN EXPANDING TISSUE

In the previous sections we have reviewed the biophysics of cell water relations and discussed spatial-temporal interrelationships in growing leaves. Understanding growth trajectories and growth velocities will now permit us to generalise Lockhart's transport equation (Equation 2, Chapter 1) to an equation for a multicellular continuum. Recognising that the tissue element is moving, we modify the flux equation to show that only water which is moving faster than the element boundary can cross the boundary to inflate the element (Silk & Wagner, 1980). An equation can then be derived to relate local growth rate and water potential. The three-dimensional treatment shows

$$R = \nabla . (\underset{\approx}{L} . \nabla\psi) \tag{3}$$

where R is the growth rate in three dimensions (equal to the relative rate of volume increase), $\underset{\approx}{L}$ is the conductivity tensor (similar to the conductance of Equation (1) but embodying different values in different directions), and $\nabla\psi$ is the gradient in water potential, the driving force for water movement. Equation (3) implies that if one knows the spatial distributions of the growth rate and of hydraulic conductivity one can solve for the distribution of water potential which is sustaining the observed growth pattern.

Evaluation of Equation (3) is still at a primitive stage. There are few data available on leaf growth distributions and equally little data on hydraulic conductivities in leaves. More is known about root growth, and it is instructive to examine the solution of Equation (3) in the maize root for which we have good data on the distribution of the growth rate (Erickson & Goodard, 1951) and good baseline data for hydraulic conductivity on the millimeter scale (Ginsburg & Ginzburg, 1970). Solutions reveal that the spatial distribution of growth sustaining water potential consists of concentric, egg-shaped isopotential regions (Fig. 3). Water potential becomes more negative with growth rate and also with distance from the water source. Thus, the point of most negative ψ is in the centre of the cross-section and at the location of the maximum growth rate. To interpret the spatial pattern correctly, we must recognise that individual tissue elements move through the pattern as they are displaced from the root apex during growth.

Fig. 3. Growth sustaining water potentials in the maize root. Contour map shows the distribution of $L\psi$ in a longisection of the growth zone of the maize root. The root is presumed to be growing in pure water ($\psi = 0$) without transpiration. Darkest region in the centre has $-L\psi > 5 \times 10^{-8} cm^2 s^{-1}$; next isopotential region shown has $4 \times 10^{-8} cm^2 s^{-1} < -L\psi < 4.5 \times 10^{-8} cm^2 s^{-1}$; and water potentials become progressively less negative with distance from the root centre and with decreasing growth rate. (Redrawn from Silk & Wagner, 1980.)

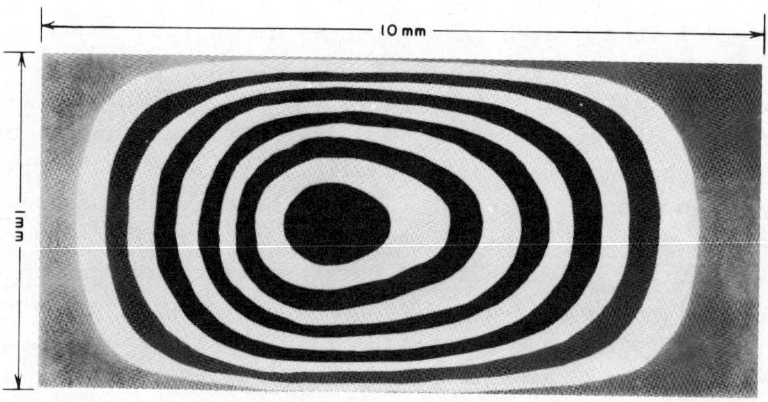

Plots of ψ versus distance from the apex show most negative ψ values in the region of maximum growth rate (4-6 mm from the tip). Plots of ψ versus time can be obtained by combining the spatial distribution shown in Fig. 3 with a growth trajectory, obtained as described for Fig. 2. The cellular particle takes 16 h for displacement from the meristematic region (1.2 mm from the root tip) to the region of most negative water potential (4 mm from the tip) and reaches maximum water potential after only three more hours of displacement and slowing expansion.

EFFECTS OF WATER STRESS

In this section we explore the implications of experiments designed to evaluate the parameters of the Lockhart equations. Most of the observations described are made on the whole growing zone without regard to spatial variation, hence, as explained in the preceding section, the results are suggestive rather than definitive. Nevertheless, our data for the maize leaf are consistent enough to provide a description of the effects of water deprivation in terms of the Lockhart model.

Expansive growth has long been recognised as being easily inhibited by water stress (Vaadia *et al*., 1961). With more definitive measurements of leaf ψ (Boyer, 1968) and methods to monitor leaf elongation minute-by-minute (Acevedo *et al*., 1971), evidences had accumulated to lead Hsiao (1973) to conclude that expansive growth is the process most sensitive to water stress. Since then a number of studies (*e.g*. Matsuda & Riazi, 1981; Michelena & Boyer, 1982) have strengthened this conclusion (Bradford & Hsiao, 1982). Results of field experiments also indicate that upon the development of water stress, restriction of leaf expansive growth is one of the very first symptoms of stress.

Role of Turgor. Reduced leaf growth under water stress has been attributed primarily to a reduction in cell turgor associated with the reduced leaf ψ (Boyer, 1968). This would explain the extreme sensitivity of the process to water stress. The relationship between cell volume and P (Tyree & Jarvis, 1982) dictates that when ψ of a rapid growing and hence quite turgid cell is reduced, P falls much more than π. Thus, a small reduction in ψ can retard or prevent growth, presumably by reducing markedly or erasing the difference between P and the yield threshold Y of Equations 1 and 2. There is yet no published study of expansive growth of leaves based on P measured directly in growing cells, although the latter measurement has been successful on pea epicotyl (Cosgrove & Steudle, 1981; Cosgrove & Cleland, 1983). Nonetheless, the role of turgor is consistent with the data of Green (1968) on Nitella, which showed virtually instant changes in growth rates in response to rapid changes in P, monitored with a micromanometer inserted in the expanding internodal cell. In an earlier study (Acevedo et al., 1971) maize leaf elongation exhibited short-term response patterns to water stress similar to those of Nitella. The results were interpreted in terms of turgor changes but only ψ of the portion of the leaf above the growing zone was measured. However, more recently, Michelena & Boyer (1982) reported data showing that as water stress developed, maize leaf elongation slowed and then stopped while P of the growing zone remained apparently unchanged. An apparent lack of turgor change was also observed by Matsuda & Riazi (1981) in the growing zone of barley leaves as growth was inhibited by water stress.

Whether the apparent lack of dependence of expansion on turgor could be caused in some cases by an artefact of measurement was investigated by Hsiao & Jing (1985). They obtained several lines of evidence indicating that wall of the cells in the growing zone probably yielded or relaxed when P exceeded Y during the time of measurement for ψ in a thermocouple psychrometer at physiological temperatures. Consequently the estimated P of samples at high turgor all attained a value similar to Y, leading to an apparently constant P for leaves of different ψ and widely different rates of elongation. Wall relaxation after excision was also indicated by the data of Cosgrove et al. (1984), obtained on growing pea stems with a pressure microprobe. Hsiao and Jing found that determining ψ quickly in the cold using the Shardakov dye method (Slavik, 1974) prevented the apparent wall relaxation. They then reexamined the relationship between growth and P in plants under water stress.

The experiments were conducted on maize seedlings

growing in soil in a growth chamber. Water was withheld from the plant pots while elongation of the fifth leaf (counting from the base) was continuously monitored with a postion transducer (Hsiao et al., 1970). When elongation of the leaf slowed to a desired rate because of water stress, the plant was excised and a 50 mm basal segment (growing zone) of the leaf was quickly sampled in a humidity chamber. Water potential was measured by the Shardakov method on one sample at 5°; π was measured on another, after freezing and thawing, using an isopiestic thermocouple psychrometer (Boyer & Knipling, 1965). The difference between ψ and π was taken as the estimate of P. One set of results are shown in Fig. 4. As the elongation slowed down due to water

Fig. 4. Elongation rate of the fifth leaf of maize (about 30 cm long) in relation to estimated P in the growing zone. Elongation rate was reduced from the maximum by fast development of water stress. Data were fitted by linear regression.

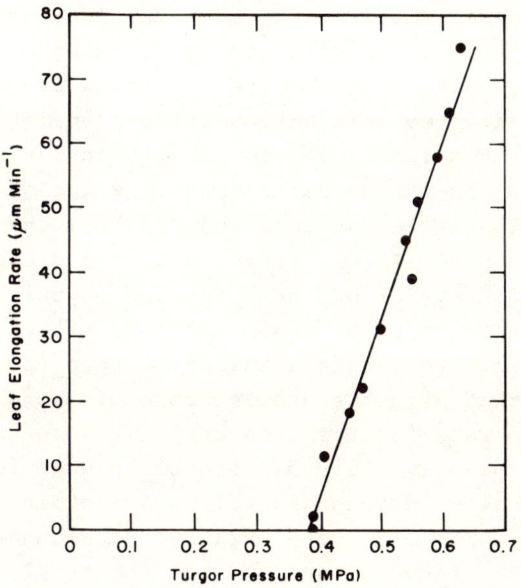

depletion, P of the growing zone dropped concomitantly. This is consistent with the view that p reduction is the initial cause of retardation in expensive growth under water stress and is in direct contrast to the apparent lack of dependence of growth on P when ψ was determined at a physiological temperature. The intercept of the line fitted to the growth rate vs. P data represents Y of the Lockhart model (Equation 2) and the slope of the line represents ϕ. Fig. 4 shows that Y for maize seedling leaf is approximately 0.4 MPa and that the dependence of elongation of P above Y is quite steep. A reduction of 0.25 MPa in P was sufficient to reduce elongation rate from its maximum to zero. The relationship between elongation rate and growing zone ψ was also approximately linear (data not shown). Elongation was reduced from maximum to zero when ψ was reduced from -0.55 to -0.9 MPa.

Osmotic Adjustment. As water deficit develops, the growing zone of various plant organs is known to undergo substantial osmotic adjustment (Meyer & Boyer, 1972; Greacen & Oh, 1972; Barlow *et al*., 1980). Similar to the finding of Michelena & Boyer (1982), data on π obtained in the above and similar experiments indicated some accumulation of osmoticum in the growing zone of maize leaves as elongation was retarded by water stress. In these experiments, water depletion was rapid and elongation rate was reduced from the maximum to zero in about 5 h. It was reasoned that a slower rate of water depletion may allow more time for osmotic adjustment. By manipulating conditions and plant competition for water, it was possible to lengthen this time span to about 19 h. Fig. 5 shows that π in the growing zone was more negative at any elongation rate for the slow drying plants than for the fast drying plants. This difference in osmoticum content enables the slow drying plants to maintain turgor and elongation in spite of a more negative ψ. Elongation was not stopped until tissue ψ dropped to about -1.1 MPa, in contrast to stoppage at -0.9 MPa for the fast drying plants (Jing & Hsiao, unpublished). These results support the concept, embedded in Equations 1a, of osmotic adjustment (lower or more negative π_i) for maintaining growth.

Fig. 5. Elongation rate of the fifth leaf of maize in relation to π in the growing zone as affected by the speed of water stress development. Elongation rate decreased from the maximum to zero in 5 h for the fast drying plants and in 19 h for the slow drying plants. After excision, samples were immediately frozen and thawed for the determination of π. Data were fitted by linear regression.

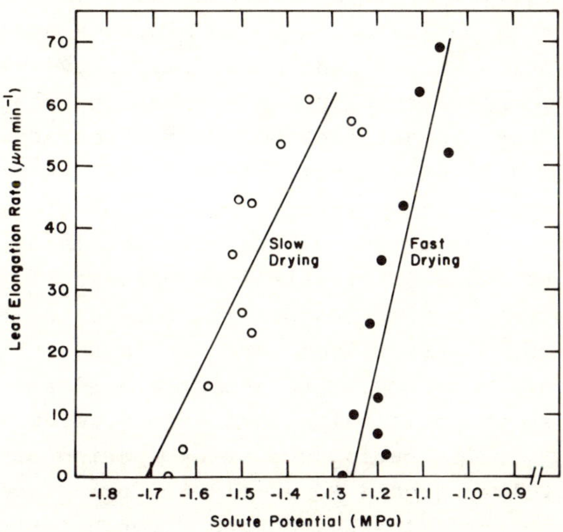

Shifts in Extensibility and Yield Threshold

Direct data indicating rapid shifts in volumetric wall extensibility ϕ and Y were obtained for Nitella undergoing changes in P and water status (Green, 1968). For higher plants, earlier there were indirect indications of shifts in wall yielding properties as water status varies (Acevedo et al., 1971; Meyer & Boyer, 1972; Bradford & Hsiao, 1982; Tyree & Jarvis, 1982). More recently, long-term drought acclimation was shown to have altered Y and possibly ϕ in leaves of sunflower in such a way that growth was slower at any given P (Mathews et al., 1984). Jing & Hsiao (unpublished) have had a limited success in following P and estimating some other parameters of Equation 1, as ψ of the rooting medium for maize was changed in a stepwise fashion. A sample of the data are shown in Fig. 6. As the rooting medium was changed from nutrient solution to -0.25 MPa polyethylene glycol (PEG) 6,000, P and ψ in the growing zone dropped approximately 0.25 MPa within minutes, and elongation stopped. This behaviour is in full agreement within what may be predicted from the relationship in Fig. 4. Initially π remained about the same, and then declined over a period of 30 to 40 min to a steady level. During that period, there was a partial recovery in growth and a nearly complete recovery in P. When the rooting medium was changed back to nutrient solution after about 3 h, P increased to a level slightly higher than that existing before the imposition of water stress, but elongation rate recovered only to about 80% of the original value.

Growth was substantially suppressed in the later half of the stress period in spite of the restoration of P. After the roots were placed back in nutrient solution, growth still did not recover to the original rate although P was higher than the original. These facts point strongly to changes in ϕ or Y during the brief period of stress.

Van Volkenburgh & Boyer (1985) found recently with maize that water stress reduced proton extrusion and acidification of the leaf surface in the rapidly growing zone. Their data are consistent with the view that acidification promotes wall loosening and that the slower growth at a given turgor in the stressed plant may be attributed to inhibition of proton extrusion.

The wall yielding characteristics in relation to turgor can also be related to the geometry of the wall and the cell. In the Lockhart model, the fundamental force for growth is not pressure but force per unit of cross sectional area of the wall, acting in the direction of expansion (Lockhart, 1965). Customarily, to avoid dealing with complex wall geometry in calculating force on the wall, Equations 1 and 2, based on pressure, are used instead of equations

Fig. 6. Changes in elongation rate of the fifth leaf of maize and in growing zone ψ, π, and P effected by stepwise changes in ψ of the culture solution. At time zero, the rooting medium of a modified 1/6 strength Hoagland solution was changed to one with polyethylene glycol 6000 added to yield ψ = -0.25 MPa. The medium was changed back to the nutrient solution after 3 h 10 min. The plants were selected for nearly identical elongation rates without water stress. Each data point represents the elongation and water potential measurements on the same leaf.

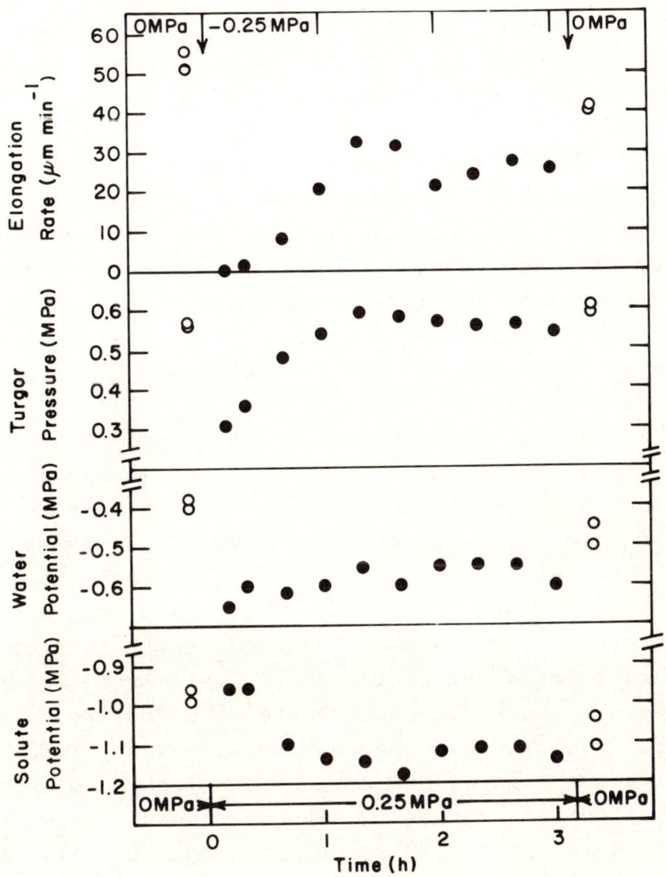

based on force. Because geometric effects are incorporated into ϕ and Y of Equations 1 and 2, they are sensitive to changes in cellular and wall dimensions. For example, if the wall thickens but otherwise remains the same, ϕ would be reduced and Y would be increased. Similarly, ϕ would decrease and Y would increase if the cell cross sectional area normal to the direction of expansion is reduced while the thickness and intrinsic yielding properties of the wall remain the same. These inherent limitations need to be kept in mind when using Equations 1 and 2. It would appear that without detailed knowledge about changes in cell and wall geometry, data on ϕ and Y based on P measurements, while useful in many ways, cannot be used to prove whether water stress affects the intrinsic yielding characteristics of the cell wall.

In the studies discussed here, as well as in most others, measurements were taken over a wide region of the growing zone and spatial details were ignored. The composite values obtained can be misleading, as stated earlier. The conclusions and interpretation made in this chapter may need alteration once the effects of water stress on spatial patterns of growth parameters are known.

Interdependence of Growth Parameters

Although this chapter has emphasised the more physical aspects of leaf growth, it was pointed out earlier that metabolism is clearly involved. This involvement is not only confined to wall synthesis and viscoelastic behaviour, but also in the regulation of osmoticum and hydraulic conductance. To maintain a given growth rate or to adjust the rate for a new environment, the metabolic and physical parameters must be orchestrated and co-ordinated through various regulatory control loops. Very little is known about these co-ordinations. Nonetheless, to emphasise the integrated nature of growth responses to water stress, Bradford & Hsiao (1982) examined conceptually the likely interdependence of the underlying parameters. A modified version of their diagram is presented as Fig. 7. The instantaneous growth rate G is depicted as a function of P, L and wall yielding characteristic, Y′, which incorporates ϕ and Y, as our knowledge on these two parameters is too limited to warrant treating them separately. In mathematics, total derivative is the sum of the partials. Fig. 7 shows that G results from or is the sum of its partials or components (G_i , where i = 1,2,...). When considering any partial functions, such as G_i = f.(Y′), it is implied that all other variables other than the independent variable Y′ are held constant.

Fig. 7. Interdependence of expansive growth and underlying parameters. G is expensive growth; M is general metabolism (substrates, energy); Y′ is wall yielding characteristic (includes volumetric extensibility and yield threshold); W is cell wall synthesis and metabolism; O is the synthesis of other macromolecules such as proteins and membranes; H is hormonal effects; and L is tissue hydraulic conductance. Each parameter is taken as a function of other parameters, e.g. G = f (P, Y′, L). Partial or component functions are denoted by subscript i. Converging heavy arrows indicate the "summing" of the partials to make the whole. Light arrows indicate the whole in turn serves as the variable on which the partial of another function depends. Loops are labelled I, II, III, and IV for ready reference. See text for further explanation.

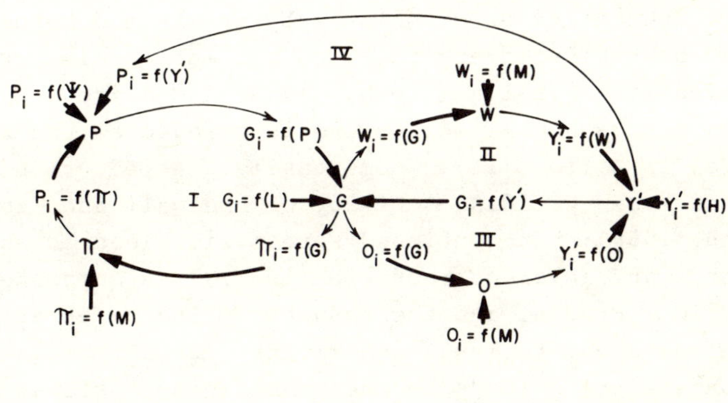

For loop I of the scheme (Fig. 7), G is partly a function of the independent variable P, which, in turn, is a function of π or osmolarity. Further on, π is dependent partly on general metabolism (M) and partly on G, hence completing a loop consisting of G, P and π. The dependence of π on G results from volume dilution due to expansion. Loop II deals with wall yielding and the synthesis and metabolism of cell wall (W). Clearly Y′ is a function of wall metabolism and hormonal activities (H). W, in turn, must depend on G, and on M to supply the wall components. Loop III connects G, Y′, and the synthesis of macromolecules other than cell wall (O). There are some data suggesting that protein synthesis under water stress is partly a function of G (see Bradford & Hsiao, 1982). Possibly membrane synthesis is similarly dependent. It is not known whether O directly affect Y′ (as arbitrarily drawn) or indirectly influencing Y′ via W. In the latter case the loop should be drawn through G, O, and W. The last loop (IV) connects Y′, P, and G. If the wall is very extensible, P will be low due to the dilution effect of rapid growth. Thus, we have the mutual dependence of G on P, and of P on G.

With a sudden development of water stress and drop in leaf ψ, the upper leg of loop I shows that P will decrease, leading to a decrease in G. The lower leg of the loop, however, shows that the reduction in G will, in turn, reduce the dilution of solutes by growth and hence leading to a partial recovery in P and resumption in growth, as observed with Nitella (Green, 1968) and maize (Fig. 6). At the same time, wall metabolism is changed as the result of change in G (loop II). Momentarily, a reduction in G may enhance the yielding ability of the wall and increase Y′ (Green, 1968); but continued stress will lead to an opposite effect, an apparent reduction in Y′, as suggested by the data in Fig. 6 and the results of Sharp et al. (1979), Michelena & Boyer (1982) and Mathews et al. (1984). Thus, prolonged stress leads to the phenomenon of slower growth at a higher or equal P. The data in Fig. 6 indicates that the duration of stress required for this effect is surprisingly short - 2 h or less!

Behaviour in the Field

What happens to expansive growth in the field under mild water stress? In previous field studies P was usually estimated in mature leaf or mature portion of the leaf instead of the growing zone. The data showed important interplays between effects of temperature and water status on growing through a duirual cycle and durial osmotic adjustment (Acevedo *et al.*, 1979). Recently we monitored momentary elongtion rate, growing zone water potential and components, and temperature in a sorghum crop growing outdoors in Davis, California. Data (Fig. 8) were collected from a well irrigated control plot and a plot which had its irrigation cut off 12 days earlier (unirrigated). Growing zone ψ was slightly lower in the unirrigated plants, especially around midday. Turgor of the growing region, however, was similar for the irrigated and unirrigated plants for most of the day. Yet, except for the early evening, leaf elongation was slower for the unirrigated plants. Hence, the phenomenon of slower growth for equal P is also observed in the field under mild water stress. Fig. 8 also shows that other than the difference between the irrigated and unirrigated treatments, growth rate appeared to be more related to growing zone temperature than to P.

Comparing the period between 9.5 and 10.5 h with that between 15 and 17.5 h, it is seen that tissue temperature was about the same for the two periods while P was lower for the later period. But elongation was faster in the later period! Apparently the tissue was more extensible or had a lower yield threshold in the later period. These results point to the possibility of a duirnal variation in wall yielding characteristics not linked to water status or temperature.

Fig. 8. Duirnal trends in leaf elongation rate, tissue temperature, ψ, and P of the growing zone in a sorghum crop 42 days after planting in the field. Water had been withheld from the unirrigated plot for 12 days on the day of measurements (Aug. 10, 1983). Temperature of the growing zone was determined with a fine-wire thermocouple. Each data point represents the elongation and water potential measurements on the same leaf.

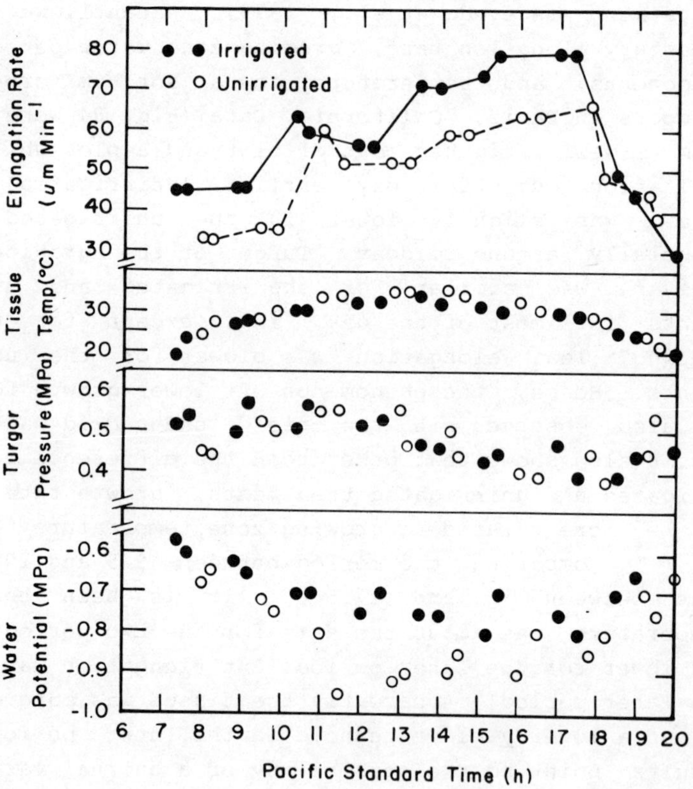

QUESTIONS FOR FUTURE RESEARCH

The maize leaf data, taken on the whole growing region, suggest that the short term effect of water stress is growth diminution produced by a drop in cell turgor. Turgor is restored in the long term by osmotic adjustment, but growth remains inhibited because the cells have become less extensible. To verify this hypothesis, we need to carry out similar studies with good spatial resolution. Effects of water stress on the spatial distribution of growth rate, turgor pressure, and osmotic potential must be determined. Only then can we examine rigorously their interactions and gain a more quantitative understanding of the growth responses to water stress. Pattern of local solute production rate should be derived from distribution of the growth rate and osmotic potential to clarify the extent to which osmotic adjustment occurs as a result of diminished growth and as a result of changes in solute importation or generation.

The physiology of the growth rate distribution in roots and leaves (Fig. 2) also remains to be elucidated. We do not understand why the growth rate increases for a long period and then declines abruptly during the rapid displacement through the distal region of the growth zone. The theory of growth sustaining ψ needs to be tested and refined. To date, only the root of *Zea mays* has been examined. Data on leaf spatial distributions (on the millimeter scale) of growth rate, hydraulic conductance and water potential are needed. The model of Lockhart needs to be improved. Means need to be developed to account for possible changes in cell and wall geometry effected by water stress. Because any cell is surrounded by neighbouring cells in a leaf, net force on a particular cell is not given solely by its own turgor pressure acting on its wall. Now pressures from adjacent cells also act and the forces experienced by a wall depends on its position relative to other cells. More work needs to be done to understand the interactions of growth forces in the cellular continuum which comprises the leaf.

REFERENCES

Acevedo, E., Hsiao, T.C. & Henderson, D.W. (1971). Immediate and subsequent growth responses of maize leaves to changes in water status. Plant Physiol. 48, 631-6.

Acevedo, E., Fereres, E., Hsiao, T.C. & Henderson, D.W. (1979). Diurnal growth trends, water potential, and osmotic adjustment of maize and sorghum leaves in the field. Plant Physiol., 64, 476-80.

Barlow, E.W.R., Munns, R.E. & Brady, C.J. (1980). Drought responses of apical meristems. In Adaptation of Plants to Water and High Temperature Stress, eds. N.C. Turner and P.J. Kramer, pp. 191-205. New York: Wiley and Sons.

Boffey, S.A., Selden, G. & Leech, R.M. (1980). Influence of cell age on chlorophyll formation in light-grown and etiolated wheat seedlings. Plant Physiol., 65, 680-4.

Boyer, J.S. (1968). Relationship of water potential to growth of leaves. Plant Physiol. 43, 1056-62.

Boyer, J.S. & Knipling, E.B. (1965). Isopiestic technique for measuring leaf water potential with a thermocouple psycrometer. Proc. Nat. Acad. Sci. USA., 54, 1044-51.

Bradford, K.J. & Hsiao, T.C. (1982). Physiological responses to moderate water stress. In Encyclopedia of Plant Physiology, New Series, Volume 12B. Physiological Plant Ecology II, Water Relations and Carbon Assimilation, eds. O.L. Lange, P.S. Nobel, C.B. Osmond and H. Ziegler, pp. 263-323. Berlin, New York: Springer-Verlag.

Cosgrove, D. & Steudle, E. (1981). Water relations of growing pea epicotyl segments. Planta, 153, 343-50.

Cosgrove, D.J. & Cleland, R.E. (1983). Osmotic properties of pea internodes in relation to growth and auxin action. Plant Physiol. 72, 332-8.

Cosgrove, D.J., Van Volkenburgh, E., & Cleland, R.E. (1984). Stress relaxation of cell walls and the yield threshold for growth: Demonstration and measurement by micro-pressure probe and psychrometer techniques. Planta 162, 46-54.

Coster, C. (1927). Die täglichen Schwankungen des Längenzuwachses in den Tropen. Rec. Trav. Bot. Neer., 24, 257-305.

Erickson, R.O. & Goodard, D.R. (1951). An analysis of root growth in cellular and biochemical terms. Growth symp., 10, 89-116.

Erickson, R.O. & Michelini, F.J. (1957). The plastochron index. Amer. J. Bot. 44, 297-305.

Ginsburg, H. & Ginzburg, B.Z. (1970). Radial water and solute flows in roots of Zea mays. J. Exp. Bot. 21, 580-92.

Greacen, E.L. & Oh, J.S. (1972). Physics of root growth. Nature 235, 24-5.

Green, P.B. (1968). Growth physics in Nitella: a method for continuous in vivo analysis of extensibility based on a micromanometer technique for turgor pressure. Plant Physiol., 43, 1169-84.

Hsiao, T.C. (1973). Plant responses to water stress. Ann. Rev. Plant Physiol., 24, 519-70.

Hsiao, T.C., Acevedo, E. & Henderson, D.W. (1970). Maize leaf elongation: continuous measurements and close dependence on plant water status. Science 168, 590-1.

Hsiao, T.C. & Jing, J. (1985). Water stress and expansive growth of maize leaves: role of turgor pressure and errors caused by apparent wall relaxation. (In preparation.)

Labavitch, J.M. (1981). Cell wall turnover in plant development. Ann. Rev. Plant Physiol., 32, 385-406.

Lockhart, J.A. (1965). Cell extension. In Plant Biochemistry, eds. J. Bonner and J.E. Varner, pp. 826-49. New York: Academic Press.

Matsuda, K. & Riazi, A. (1981). Stress-induced osmotic adjustment in growing regions of barley leaves. Plant Physiol., 68, 571-6.

Matthews, M.A., Van Volkenburgh, E. & Boyer, J.S. (1984). Acclimation of leaf growth to low water potentials in sunflower. Plant, Cell and Environment 7, 199-206.

Meyer, R.F. & Boyer, J.S. (1972). Sensitivity of cell division and cell elongation to low water potentials in soybean hypocotyls. Planta 108, 77-87.

Michelena, V.AS. & Boyer, J.S. (1982). Complete turgor maintenance at low water potentials in the elongating region of maize leaves. Plant Physiol., 69, 1145-49.

Ray, P.M., Green, P.B. & Cleland, R. (1972). Role of turgor in plant cell growth. Nature 239, 163-4.

Richards, O.W. & Kavanagh, A.J. (1943). The analysis of the relative growth gradients and changing form of growing organisms: illustrated by the tobacco leaf. Am. Nat., 77, 385-99.

Sharp, R.E. Osonubi, O., Wood, W.A. & Davis, W.J. (1979). A simple instrument for measuring leaf extension in grasses, and its application in the study of the effects of water stress on maize and sorghum. Ann. Bot., 44, 35-45.

Silk, W.K. (1980). Plastochron indices in cantaloupe grown on an irrigation line source. Bot. Gaz. 141, 73-8.

Silk, W.K. (1983). Kinematic analysis of leaf expansion. In The Growth and Functioning of Leaves, eds. J.E. Dale and P.L. Milthorpe, pp. 89-108. Cambridge:

Cambridge University Press.

Silk, W.K. (1984). Quantitative descriptions of development. Ann. Rev. Plant Physiol., 35, 479-518.

Silk, W.K. & Erickson, R.O. (1979). Kinematics of plant growth. J. Theor. Biol., 76, 481-501.

Silk, W.K. & Wagner, K.K. (1980). Growth-sustaining water potential distributions in the primary corn root. Plant Physiol., 66, 859-63.

Slavik, B. (1974). Methods of studying Plant Water Relations. Ecological Studies, Volume 9, Analysis and Synthesis, eds. J. Jocobs, O. Lange, J.S. Olson, and W. Wieser. New York: Springer-Verlag.

Tyree, M.T. & Jarvis, P.G. (1982). Water in tissues and cells. In Encyclopedia of Plant Physiology, New Series, Volume 12B. Physiological Plant Ecology II, Water Relations and Carbon Assimilation, eds. O.L. Lange, P.S. Nobel, C.B. Osmond and H. Ziegler, pp. 35-77. Berlin, New York: Springer-Verlag.

Vaadia, Y., Raney, F.C., & Hagan, R.M. (1961). Plant water deficits and physiological processes. Ann. Rev. Plant Physiol., 12, 265-91.

Van Volkenburgh, E. & Boyer, J.S. (1985). Inhibitory effects of water deficit on maize leaf elongation. Plant Physiol. 77, 190-4.

LEAF GROWTH AND WATER DEFICITS: BIOCHEMISTRY IN RELATION TO BIOPHYSICS

D.W. Lawlor and J.E. Leach
Physiology and Environmental Physics Department,
Rothamsted Experimental Station, Harpenden,
Hertfordshire, AL5 2JQ, United Kingdom.

INTRODUCTION

Size and composition of leaves and their efficiency in photosynthesis and other metabolic processes are mainly determined by genotype but phenotypic expression is substantially affected by the plants environment, principally temperature, nutrition, irradiance and water supply (Terry *et al*., 1983). When plants lose water faster than it is transported into the roots, they suffer water stress, a decrease in the content and free energy of cellular water resulting in a decrease in the rate or total amount of plant growth in size, mass or energy or changes in the differentiation of organs. In many climates water stress, of varying magnitude and duration, occurs commonly during leaf development, decreasing productivity by slowing the rate and shortening the duration of leaf growth and increasing senescence. Consequently stressed plants have fewer, smaller leaves and therefore less total leaf area to intercept radiation (Legg *et al*., 1979). Leaf composition (*e.g*. dry mass/area, amino acid content) also changes with stress, indicating alteration to the metabolic "machinery" of the cells (Lawlor, 1983).

The greatest effects of stress occur during leaf development, when new cells and their constituent organelles, walls, *etc*. are formed. Metabolism is then very active and demand for materials and energy is large. Water stress decreases cell growth and leaf expansion, slows the uptake of nutrients and assimilation, and increases senescence. Fig. 1 illustrates the hierarchy of events during the development of leaf area. Changes in growth,

Fig. 1. Hierarchy of processes relating cell metabolism to crop leaf mass and area, much simplified. Water stress of different severity affects processes at each level of organisation.

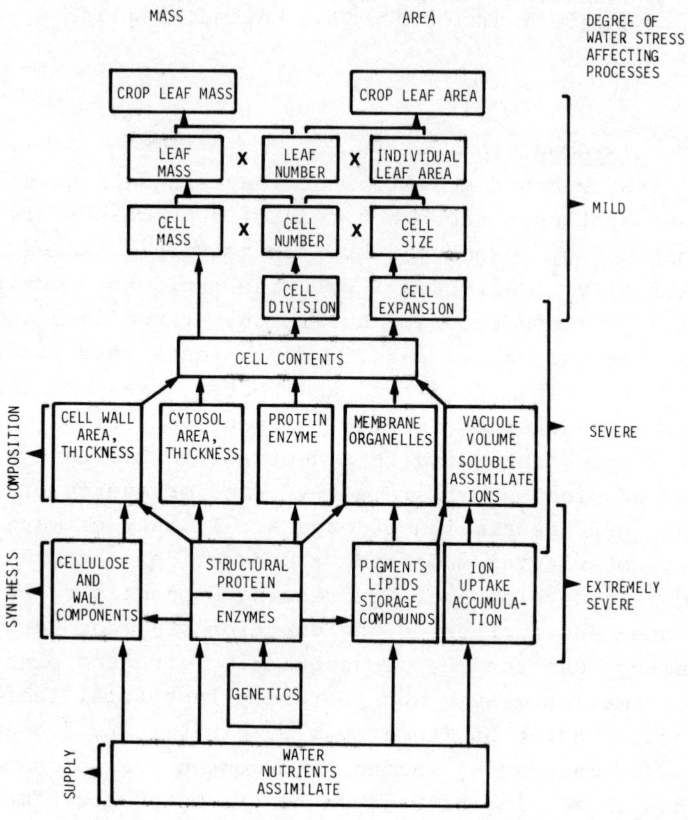

assimilation efficiency and senescence of stressed plants may be related because changes in the biophysics of the cell will affect the metabolic systems and thereby leaf function. We consider in this chapter how changes in cell biophysics, caused by water stress, may interact with metabolism and influence leaf development. Present concepts are only qualitative and in some areas speculative; quantitative models are required to understand not only effects of drought but how leaf growth in general is controlled.

CELL DEVELOPMENT AND WATER STRESS

Meristematic cells are 5 to 20 μm in diameter, with thin primary walls and few, small vacuoles scattered throughout the cytoplasm. They contain prominent nuclei, endoplasmic reticulum and polyribosomes which indicate rapid synthesis of protein. The abundant Golgi bodies and microtubules are part of the machinery of wall synthesis and deposition, which are active processes during cell division and growth (Robards, 1970). Growth and differentiation, i.e. change in form and function, require synthesis of cell components in the correct amount and sequence and their incorporation into structures, as part of a highly integrated and controlled process. Mitotic division of cells of the apical meristem gives rise to the leaf buttress, which develops into the young leaf by continued cell division. With increasing age of the leaf, cell division decreases although meristems persist in parts of the leaves of dicotyledonous plants, and at the base of leaves in monocotyledonous plants, so leaf cells are of different ages (Lyndon, 1983). The cell population is heterogeneous with some cells photosynthetic and others active in storage or transport, which differ in metabolism. Little is known of the effects of stress on particular cell types and in this paper we consider a generalised mesophyll cell. During rapid expansion of cells, which is responsible for most of the increase in leaf size, water accumulates in large central vacuoles. Expansion is accompanied by active synthesis of components of cell walls, membranes and organelles, e.g. proplastids, which develop into chloroplasts in the light. Photosynthesis develops early in the expansion of leaves so that regulation of cell metabolism involves photosynthetic and respiratory energy and material fluxes, together with the formation of cell structure and synthesis and control of osmotic constituents.

BIOPHYSICS OF CELL WATER BALANCE, ELASTICITY AND GROWTH

Concepts of cellular biophysics (Tyree & Jarvis, 1982) can be used to establish the principles of cell growth and to identify how metabolic processes may be affected by stress. Much of the information that follows relates to mature cells and, although applicable to developing cells, may not describe their condition fully.

Cell Water Balance

The chemical potential of water (free energy per mole) μ_w, in a multicomponent, open system and the effects of pressure, P, absolute temperature, T, and position in the gravitational field, h, are related by:-

$$\mu_w - \mu_w^o = RT \ln a_w + P\overline{V}_w + ghm_w \qquad (1)$$

where μ_w^o is the μ_w of pure water under standard conditions of temperature and pressure and position in the gravitation field (a reference usually defined as zero potential), R is the gas constant (8.314 J mol^{-1} K^{-1}), g is acceleration due to gravity, m_w is the mole mass of water and a_w is the activity of water, defined as the mole fraction; $a_w = n_w/(n_w + \Sigma_i n_i)$ where n_w and n_i are the moles of water and solutes respectively. $\overline{V}_w$ is the partial molar volume of water (ca. 18 cm^3 mol^{-1}). Activity of water is measured by the decrease in water vapour pressure (e) in equilibrium with a solution compared to that of pure water (e_o); $a_w = e/e_o$. Increasing P and h above the reference increases μ_w, i.e. it is positive compared to μ_w^o. Addition of solutes (e.g. sugars, salts) to pure water decreases the mole fraction and therefore activity of water and μ_w is negative with respect to μ_w^o. Solid surfaces, for example in porous media (cell walls) and in capillaries (xylem vessels) bind water molecules, causing an additional matric potential. Its magnitude depends on the water content and it is difficult to distinguish from the effects of solutes. Only at very low water contents, which are lethal to actively growing plants, or in water movement, is matric potential a major component of μ_w in mesophyll cells but it may be important for maintaining tissue or cell water content in apical meristems or in seeds during severe stress; it is not considered further here. Also the effect of gravity is unimportant within individual cells and will be neglected (h=0).

By expressing the terms in equation (1) per unit of partial molar volume, the water potential, $\psi = \mu_w - \mu_w^o / \overline{V}_w$, the osmotic pressure, $\pi = RT \ln a_w / \overline{V}_w$, and the pressure potential, $P = P\,\overline{V}_w / \overline{V}_w$, can be defined and equation (1) becomes:-

$$\psi = \pi + P \qquad (2)$$

The energy units of chemical potential (J mol^{-1}) become Pascal (Pa) pressure units, as 1 J = Kg $m^2 s^{-1}$ and J mol^{-1}/m^3 mol = Kg m^{-1} s^{-1} or Newton m^{-2} = Pa. Therefore the potential of water and its behaviour in chemical and physical processes is determined by solute activity (in ideal solutions equivalent to concentration) and the applied pressure. In plant cells the terms in equation (2) are identified with cell water potential, (ψ), osmotic potential (π), and internal hydrostatic pressure (P), called turgor pressure.

Plant cells are osmometers, with a semi-permeable lipid-protein membrane lining the cellulose wall and enclosing the cell contents, including all metabolic machinery. Accumulated solutes (organic molecules and ions) produce the cellular osmotic potential, π. When a cell, not saturated with water, is placed in pure water, water molecules diffuse down the chemical potential gradient, caused by π, across the membrane into the cell thus increasing its volume, V. The force or stress increases on the wall, which deforms and the tension between wall components increases. Stress/strain relationships of cells are determined by the cell wall, not by cytoplasmic membranes, which rupture under small tension. The resistance of the cell wall to expansion produces the turgor pressure which increases until equal to π, when the cell water potential, ψ, at the cell surface outside the plasmlemma, is zero. To maintain an influx of water into the cell and maximise water content, ψ must be lower (i.e. more negative) than the potential of water in the environment, ψ_e. Bathing a fully turgid cell in a solution equal in potential to π causes water efflux and equilibrium is reached when P is zero.

The flux of water, F, between the cell and its environment is determined by the difference of ψ between the cell surface and the medium ($\Delta\psi = \psi - \psi_e$) and the hydraulic conductivity of the cell surface, L; i.e. $F = \Delta\psi / L$. However ψ does not show the contributions of π and P to the state of water inside the cell to which metabolism is related and ψ does not directly control the biochemical processes. We consider that biochemical processes are "adjusted" to a particular combination and range of internal conditions for normal function. In this view, water stress decreases metabolic efficiency and therefore growth because changes in π and P exceed the capacity of metabolism to adjust. Some cellular processes appear to be more sensitive to π (e.g. enzyme reactions) and others to P (e.g. cell expansion). Differences in response of cells from different species to water stress may depend on the relative importance of π and P in metabolic reactions. Species differ in the relationship of π and P, as will be discussed later.

However the effects of solutes are not only via the effect on water potential. Solutes displace water molecules from around macromolecules disrupting their structure and bonding characteristics; solutes such as proline are thought to structure water and therefore maintain macromolecular activity. Charged solutes also interact with the bonding in proteins and membranes, and generally decrease metabolic efficiency. Turgor acts mainly via ψ, as the pressure in the range experienced by plant cells is too small to alter water structure. It is, however, responsible for cell growth and wall expansion and thereby influences metabolism.

Osmotic potential of cells is a most important determinant of cell biophysics and metabolism. Rates of synthesis and consumption of organic solutes in metabolism, and their export and import, determine osmotic potential and the type and amount of solutes accumulated. Ion uptake and accumulation depend on the amount and activity of the transport systems and availability of energy to exchange ions with the environment. Solute accumulation is therefore directly dependent on metabolism.

The magnitude of π and the contribution of different solutes to it, depends on species and growing conditions. Ions, mainly K^+ and Na^+ (the latter particularly in halophytes) and sugars contribute most to π, with potassium salts of organic acids of general importance and amino acids of importance in some species. Water stressed plants may accumulate proline and other amino acids and osmotically stressed algae accumulate polyols (polyhydric alcohols) such as mannitol and glycerol (see Wyn-Jones & Gorham, 1983).

Within a cell π is the same between compartments (e.g. vacuole and organelles) as membranes are not strong enough to maintain a gradient of P between them. the osmotica giving π may, however, differ between compartments. In the cytosol K^+, polyols, amino acids and quaternary amino compounds (e.g. glycine betaine), are important. These solutes cause less disruption of cytosolic functions than other solutes (e.g. Na^+, organic ions) and are compatible with protein and subcellular structure and maintain cellular function at high concentration.

Solutes, enzymes and other proteins, and membranes in cells are very heterogeneous, differing in electrical charge, molecular size and bonding properties through which they interact. Structure of catalytic sites and the activity of enzymes depend on solute composition and concentration. A decrease in cell water content alters the spatial distribution of organelles and enzymes and increases solute concentration, changing the osmotic and ionic environment and the structure of water around macromolecules. Also concentrations of metabolic control molecules may alter, affecting rates of metabolism. Changes in osmotic/ionic concentration are more important than turgor pressure changes which, in the range experienced by plant cells (0 to 5 MPa), do not affect metabolic reactions significantly.

Cell metabolism is probably optimised to a narrow range of osmotic and ionic composition and concentration in the cytosol, although little is known of relationships *in vivo*. Cytosolic functions are protected from unfavourable conditions, *e.g.* salinity, by sequestration of ions (particularly Na^+) and storage of organic acids and sugars in vacuoles. Ion pumps and metabolite translocators regulate composition by expending energy, *e.g.* ATP for Na^+ and K^+ ATPase activity. Isolated enzymes and organelles *in vitro* require control of pH, ionic composition and ionic strength for efficient function. Osmotic potential *per se* is less important for enzymes than for organelles, with semi-permeable bounding membranes, which are sensitive to changes in external osmotic potential but may adjust their composition with time. If the capacity of the control and regulatory mechanisms is exceeded by the input of ions or there are rapid changes in osmotic potential, then metabolic efficiency falls and damage may result.

Osmotic Adjustment

Plants of saline or dry habitats have lower π (-2.0 to -4.0 MPa) than mesophytes (-0.8 to -1.5 MPa), an advantage in adverse environments for low π enables turgor to be maintained and growth (discussed shortly) to continue. As turgor decreases ψ falls without a large decrease in V and π. However once turgor reverts to zero, further substantial decrease in ψ results only from a large decrease in V, which concentrates the cell osmotic contents in proportion to volume change. This decreases π and ψ and increases the gradient of ψ between soil and leaf. Such a mechanism may maintain water flux into plants during very short term stress, but operates only when turgor is lost, which (amongst other things) stops expansion growth and causes stomata to close, slowing photosynthesis and increasing photorespiration (Lawlor, 1979).

Many mesophytes, particularly resurrection plants, mosses *etc*. can survive low cell water content for long periods but leaves of mesophytic plants senesce rapidly. Meristematic tissues of mesophytes, *e.g*. resting apices, are enclosed in protective coverings which substantially decrease evaporation, and their position close to the roots or xylem may minimise decrease in ψ. However the extent to which their ψ follows the rest of the plant, particularly under severe stress is not clear. Meristematic or young cells are more resistent to stress than older cells, possibly because their protein synthesising systems are active and this enables metabolism to respond quickly to changing conditions.

Osmotic adjustment provides a mechanism to maintain P and V whilst allowing ψ to decrease. The magnitude of decrease in π varies with species from 0 to 1 MPa and may occur in hours or, more usually, days to weeks. Metabolically compatible solutes are synthesised and accumulate, probably in the cytosol; this demands materials, *e.g*. assimilates and nitrogenous compounds, which otherwise could be used for growth. Ion accumulation and maintenance of ion gradients require additional energy although for an equivalent change in π, ion accumulation uses only 10% of the energy of *de novo* synthesis of solutes. Ion concentration appears to reach an upper limit, dependent on species, beyond which the cell cannot survive; this limit is reached when ion concentration exceeds the capacity of ion pumps and damages membranes and macromolecules. In the cytosol, compatibility between solutes and macromolecules is also limited and enzyme activity may be inhibited at high concentration, *e.g*. Larkum & Wyn-Jones (1979) observed that concentrations of glycine betaine in excess of 500 osmol m^{-3} slowed the rate of enzyme reactions. Developing and expanding cells accumulate solutes under stress but accumulation of ions may contribute less than in mature cells (Delane *et al.*, 1982); the relative importance of particular ions and solutes in osmotic adjustment in tissues of different ages and their effects on metabolic processes is not known.

Adjustment of π may maintain turgor, but metabolic functions, *e.g.* growth or stomatal opening, may be slowed, probably because cytosol composition and ionic concentration during osmotic adjustment are sufficiently different from the non-stressed state to inhibit metabolism. Osmotic adjustment occurs in many families and genera but it is not always consistently observed, *e.g.* in wheats, *Triticum dicoccum* adjusts but *T. aestivum* does not (Turner & Jones, 1980), suggesting that it is not a general mechanism for overcoming drought. There may be considerable differences between species in response to stress, depending on their eco-physiological adaptations, and in the types of solutes employed.

Cell Elasticity

Cell walls are imperfectly elastic, *i.e.* will almost return to their original form when an applied stress, such as a change in P, is removed. Elasticity depends on bonding between components of the wall. Primary walls are of a "multinet" construction with cellulose fibrils orientated at an angle to the cell axis and linked together by other wall components, hemicellulose, pectin, protein, *etc.* (Albersheim, 1976).

The relationship between V, P and π depends on the bulk modulus of elasticity, ϵ, of the cell (see Tyree & Jarvis, 1982):-

$$\epsilon = (dP/dV).V \qquad (3)$$

Small ϵ signifies elastic cells, large ϵ rigid ones. Measurements of ϵ, made with a pressure probe on individual cells, show that ϵ increases with P and that small, expanding cells have smaller ϵ than large, mature cells (Zimmerman, 1977). Bulk elastic modulus is not constant, but changes with the applied turgor, e.g. increasing from ca. 1 to 200 MPa as P increases from 0 to 1.5 MPa in Populus leaves (Tyree & Jarvis, 1982). The changes in ϵ may arise because the contribution to the elasticity from parts of the wall, having different stress/strain relationships due to bonding between particular components, changes under increasing turgor. Weak bonds betwen components might allow the wall to extend under small applied force but once these bonds are deformed, other, stronger linkages must be changed, requiring a greater applied force.

Growing and developing cells of small volume have lower ϵ (i.e. are more elastic) than larger, mature cells (Zimmerman, 1977). Probably wall structure, which confers plasticity and allows growth, also makes the wall elastic, but as growth stops and the cells mature, their walls become rigid and ϵ increases. Elasticity is greater in large cells, because (as discussed above) the wall has regions of different elasticity (Zimmerman, 1977).

The tension developed in the wall as a result of a given P depends on wall thickness, σ, and cell radius, r (Lockhart, 1965; Nobel, 1974). In isodiametric cells, the total force on the wall is $4\pi r^2 P$ acting over cross sectional area of $2\pi r\sigma$, and the force on unit cross section of wall is $s = 2rP/\sigma$. In cylindrical cells, the stress per unit sectional area of wall in the tangential direction is twice that in the longitudinal direction; both increase with r. Therefore large diameter cells with large P will experience considerable forces in the walls. Nobel (1974) calculated the tangential stress in walls of Nitella to be 60 MPa at P of 0.6 MPa, but between P of zero and 0.6 MPa walls deform by only 4% of their length. Probably cell wall and osmotic characteristics are balanced to optimise water relations with the need for materials and energy in wall formation.

The relationship between cell volume, π, ψ and P depends on the particular species, cell type and habitat in which cells grow, and on cell wall properties. Very elastic (low ϵ) or inelastic (high ϵ) cells are characteristic of mesophytes and xerophytes respectively, shown schematically in Fig. 2. Water content (expressed as relative water content, RWC = actual cell water content/fully turgid cell water content) of mature, elastic cells may change from 1.0 to 0.7 between full and zero turgor (ψ range 0 to -2.0 MPa) whereas for inelastic cells RWC may change from 1.0 to 0.85 over the same range of ψ. The slope of the relation between P and RWC (Fig. 2) is important for plant growth and survival in drought conditions. It determines the changes in turgor and cellular water which occur as a consequence of maintaining the gradient of water potential between soil and leaf. Decreasing P allows ψ to decrease substantially with smaller changes in V and π than required if concentration of cell contents were the only mechanism. It is a homeostatic

Fig. 2. Relationship between cell water content (RWC) turgor pressure (P) and water potential (ψ) for cells with different osmotic potentialso (π) and elastic moduli. Non-elastic cells (ϵ = 9.5 MPa) experience less change in water content for given change in ψ than elastic cells (ϵ = 4.5 MPa).

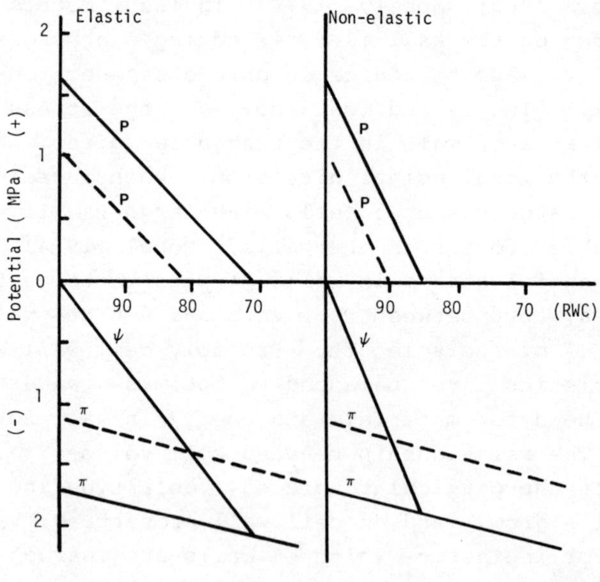

mechanism, maintaining constancy of the internal cellular environment for metabolism.

To generate low ψ with minimum change in V, a large P is required (i.e. π must be low) and the bulk elastic modulus should be large. Xerophytes can, therefore, attain a given ψ at higher V (Fig. 2), and have in general, more negative osmotic potential so that P and V are maintained at lower ψ. If cell functions are inhibited by loss of water and increased solute concentration, then inhibition in metabolism and growth would be greater for a given decrease in ψ in mesophytes than in xerophytes. Large, elastic cells experience larger change in V and π during water loss whilst maintaining a given P; plants with such cells appear to suffer severely from relatively mild drought. Thick walled non-elastic cells change ψ and P without large change in V and π. In addition, small cells with thick walls are probably better able to withstand large P generated by low (more negative) osmotic potential than larger, thin walled cells. Mature cells of plants grown under stress do have thicker walls than those grown without stress. For example, in cotton, wall thickness was 0.78 μm in stressed compared to 0.68 μm in unstressed plants (Cutler et al., 1977). Presumably wall material is deposited over a smaller area. Composition as well as thickness may be expected to change and produce walls with a different relationship of ϵ to P so that under stress conditions cells would become less elastic. However data for crops grown in the field are equivocal; there was no change with drought in rice (Cutler et al., 1980) or sunflower (Turner & Jones, 1980) but in sorghum ϵ adjusted over a long period, although this adjustment was possibly caused by changes in cell size and other characteristics as the cells aged rather than structural changes in the cell wall (Tyree & Jarvis, 1982).

Leaves grown in warmer conditions with abundant water and nitrogen have larger cells, and higher water to dry matter ratios than those grown under drought. Cotton leaf cells decreased from 37 to 32 μm in diameter and by 35% in volume as a consequence of stress (Cutler et al., 1977).

The surface area to cell volume ratio of spherical and cylindrical cells increases as the radius decreases, so with walls of constant or greater thickness and the same density the ratio of mass of wall to volume of cell water is greater in small cells. Also with a constant cytosol "thickness" the ratio of cytosol to water volume increases, and the amount of protein and nitrogen per unit area will increase with stress, as is frequently observed. Horticulturalists call plants grown without stress "unhardened" and those grown with some degree of stress "hardened". Unhardened plants are more sensitive to drought than hardened ones, but the metabolic and physiological adaptations involved are complex and have not been explored in detail.

The combination of low π, high ϵ and thick walls is an adaptation to drought. However extra material and energy is needed for wall synthesis and to maintain low π. Also cell division and growth require walls, which expand easily and which may conflict with the need for thick, rigid walls. Very low π may, in developing cells, produce large P and apply too much stress on the softer, growing walls. As dividing and developing cells grow in volume, their walls grow in area and are not strengthened mechanically; even in xerophytes they are "soft" and grow (often only in flushes) when conditions are suitable. Probably there is a range of P, and therefore π, which enables the wall to expand without over-stressing the developing structure as bonds between wall components are loosened metabolically and material is deposited.

Rate of Cell Growth

The physical factors determining cell growth are discussed extensively in Chapter 1 and will not be considered further here. Biochemical mechanisms underlying the processes of wall growth in mass and size are understood in general terms. Plastic deformation of expanding cell walls results from metabolic loosening of the bonds between polysaccharides (and proteins?) in the wall. Cellulose fibrils move relative to each other in a series of viscoelastic extensions and new material is deposited in the matrix by the process called "intussusception". The mechanism of wall loosening and extensibility (see Chapter 1) involves auxin which stimulates cell membranes to excrete a "wall loosening factor", now identified with some certainty as H^+, into the wall fabric where it regulates extensibility but not the yield threshold for growth (see Chapter 1). Possibly enzymes within the wall, regulated by H^+ concentration, catalyse the hydrolysis of polysaccharide bonds allowing wall components to move. Abscicic acid (ABA), which accumulates in young, growing leaves under stress conditions, may interfere with the H^+ exchange mechanisms and inhibit wall extension, but its role is unclear (Goodwin & Erwee, 1983). Inadequate synthesis of wall components due to impaired carbohydrate metabolism may limit wall growth, but probably only at severe stress. Experiments with dwarf and normal peas suggests that growth is regulated by soluble carbohydrate concentration, which is controlled by gibberellins; these hormones control synthesis of carbohydrate-metabolising enzymes. Soluble carbohydrates maintain π during water uptake so growth may continue despite cell enlargement and dilution of the contents. However under stress, osmoregulation (<u>i.e</u>. decrease in π) only partially maintains turgor (P) (Cleland, 1977) and therefore the rate of cell volume growth decreases as stress develops. Hormonal regulation of growth, <u>i.e</u>. cell expansion and division, is not well understood, even in unstressed tissues.

Water stress slows the rate but not the duration of cell expansion, <u>i.e</u>. metabolic events causing wall "hardening" occur at the same rate in stressed and unstressed cells. Thus, as mentioned above, mature cells, developed under stress, are smaller (Henckel, 1964) with thicker walls and a greater ratio of dry matter to volume. How this affects cell function is a matter for speculation. If cells accumulate constant amounts of solute per cell then small cells would have a lower π than large cells; continued synthesis and accumulation in a smaller volume could be a reason that stressed cells usually have more negative π. However osmotic adjustment by synthesis of particular solutes implies control of solute accumulation and production, possibly operating on metabolic processes via a turgor, or elasticity-sensing mechanism in membranes (Zimmerman, 1977). Proximity of membrane components might alter the electrical potentials on membranes, change ion binding and influence the reaction rates of ion pumps, etc., so that π is, at least, partly controlled.

For many species after leaf growth has been stopped by stress during rapid expansion, rewatering greatly increases the rate of leaf extension over many hours, compared to unstressed plants (Lawlor, 1972). This may occur because π is low and P large until π returns to normal; alternatively wall extensibility or yield threshold may change allowing greater rate of cell volume growth per unit of P. As stress develops, wall material may be synthesised and stored in the cytosol or incorporated into the wall, which cannot expand until P is greater than the yield threshold. When P increases rapidly upon relief of stress, wall synthesis may then occur by expanding the preformed wall or by drawing on preformed materials. Osmotically active solutes are either synthesised very quickly during rapid growth to maintain π, or solutes accumulated during stress maintain large P. Rapid growth after stress provides a useful, but unexploited system for the study of cell biophysical-biochemical interactions.

Cell Division

Meristems are considered to maintain cell division, even under stress which stops cell expansion; osmotic adjustment occurs in growing leaves and floral apices (Munns & Weir, 1981; Barlow et al., 1980). However severe stress does inhibit cell division, probably because turgor is too small to produce the force required for expansion to a "critical" size to allow division (Hsiao, 1973). Division might be inhibited indirectly by inadequate protein or cell wall synthesis. Longer-term effects might be hormonal or nutritional and therefore be expressed at more moderate stress.

INFLUENCE OF WATER BALANCE ON METABOLISM

Changes in turgor (P) and cell volume (V) over a much greater range than experienced by plants, probably have little direct effect on enzymes. Stresses that cause a 20% decrease in RWC may cause enzymes to precipitate or become crystalline when the normal concentration of the enzyme is very large, e.g. RuBPcarboxylase in the chloroplast. Loss of cell volume may have a greater effect on organelles and membranes. Chloroplasts and mitochondria are osmotic sacs, shrinking as V and external π decrease; their membranes contract in area in proportion to $(\text{radius})^2$ but their contents concentrate in proportion to $(\text{radius})^3$, a 20% decrease in volume decreases area by 14%; increased ion and solute concentrations may affect enzyme reactions and smaller surface area alters the spatial arrangements of the components in membranes. Mild water stress (10% loss of turgor) may decrease cell expansion by more than 50% (Cleland, 1977) and the smaller cell walls and membranes will require less material for their synthesis although as already mentioned walls may be thicker with stress. Smaller cell volume also requires less material to generate a given π. Therefore stress which slows cell expansion may cause assimilates to accumulate and, by altering the concentration of products and precursors affect particular reactions through "mass-action".

Rates of metabolic processes depend on number and turn-over time of enzyme active sites, on substrate concentration, and on the concentrations of control ions, molecules, etc. Complex interactions between these parameters control enzymes at many sites in metabolic networks and provide "feed-back" and "feed-forward" control of reaction rates and allow optimisation of material and energy fluxes at particular concentrations of substrates. An example is metabolic regulation of starch synthesis from glucose by the ADP-glucose pyrophosphorylase which is inhibited by large inorganic phosphate concentrations but activated by high 3-phosphoglyceric acid concentration (Preiss & Kosuga, 1976). Decreases in cell volume could increase inorganic phosphate concentration and inhibit starch production, so that π would be kept low by accumulated soluble sugars, but this would be a very crude control and is unlikely to be of great importance except perhaps with extreme and rapid stress.

Regulation of enzymes by ions is common; e.g. coupling factor I, the enzyme responsible for ATP synthesis in chloroplasts, has optimum activity at 1 to 3 mM Mg^{2+} but the decrease in V may increase Mg^{2+} concentration to 5 mM and inhibit photophosphorylation (Boyer & Younis, 1983). Enzymes, even from halophytes, function optimally at ionic concentrations of less than 100 mM and are protected from large ionic concentrations *in vivo* by compartmentation. Organic solutes, which maintain the activity of enzymes, accumulate in the cytosol. Energy, as ATP, and substrates, e.g. nitrogenous compounds, are prerequisites for metabolic regulation; if deficient they will limit osmotic adjustment. Little is known of the regulation of π or ion concentration; ion pumps in membranes may play a role and turgor pressure may be a control signal. Metabolic processes differ in their responses to stress and we now consider, in a simplified way, some processes of importance to cell function.

Respiration and Provision of Materials for Biosynthesis

Developing leaf cells have large requirements for energy (as ATP), reductant (as NADH) and for many carbon substrates; these are derived from respiration initially, using stored and translocated carbohydrates. Translocated amides and amino acids, from storage proteins, etc., provide nitrogen, and fats may also be derived from reserves. It is not known if the proportions of fats and proteins respired increases with stress. Provision of substrates for synthesis is of greatest importance during rapid growth, e.g. sugar phosphates are needed for cellulose synthesis, and acetyl

CoA for pigments, lipids and hormones. Amino acids are required for protein synthesis and their metabolism is linked, via glutamate, to the Krebs cycle. Respiration is greatest in dividing cells and decreases during expansion. Mild, moderate and severe stress (ψ_e in the very approximate ranges 0 to -1.5 MPa, -1.5 to -2.5 MPa and -2.5 to - 5.0 MPa respectively) decreased oxygen uptake in young wheat apices by 10, 35 and 60% (Pheloung & Barlow, 1981) but growth of leaves, and perhaps cell division, is halted by stress of -2.0 MPa. This and other evidence (Hsiao, 1973) suggests that respiration is much less sensitive to stress than leaf expansion. The decrease in respiration may not be caused by decreased substrates, as carbohydrates and amino acids often accumulate. Oxidative phosphorylation (ATP synthesis) is relatively insensitive to moderate stress, in contrast to photophosphylation, but it is decreased by severe stress for example when isolated mitochondria are osmotically stressed. ATP synthesis is essential for maintenance of cell function and synthetic processes in intermediary metabolism and impaired supply would very rapidly inhibit metabolism by slowing protein synthesis and ion pumping and would decrease growth by decreasing cell expansion and possibly cell division.

Nucleic Acid and Protein Synthesis

Both processes (see Fig. 3) are absolute requirements in dividing and differentiating cells when structures are laid down. Moderate stress probably has little effect on nucleic acid composition as it is highly regulated (Hsiao, 1973) and conserved but severe deficiency of ATP might slow nucleic acid synthesis and prevent genome replication. The amount of different RNA species in stressed tissues appears not to have been estimated. In very stressed tissues RNAase activity increases and may inhibit accumulation of RNA. In protein synthesis amino acids are activated, forming aminoacyl tRNA's which are required for peptide assembly (Fig. 3). Ribosomes on mRNA, the polyribosomes, provide the template for peptide synthesis

Fig. 3. An outline of protein synthesis. Large arrows show possible site of control by stress.

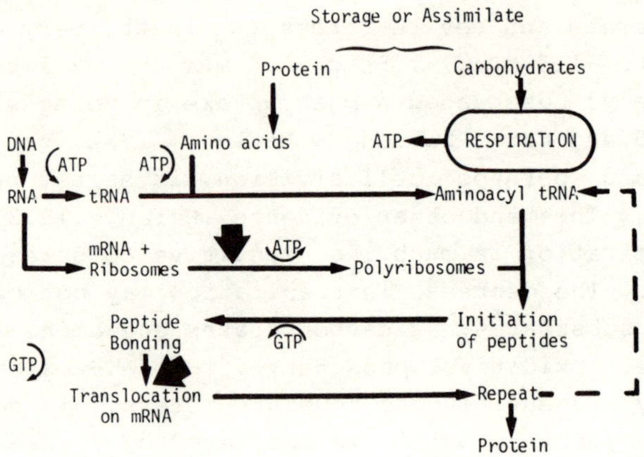

(Fig. 3) and are characteristic of cells actively making proteins. Quite mild water stress (-1.0 to -1.5 MPa) rapidly decreases the number of polyribosome in growing tissues and slows or stops protein synthesis; this is quickly reversed on rewatering (Hsiao, 1973). The percentage decrease in polyribosome number is greater than the decreases in respiration in wheat apices (Pheloung & Barlow, 1981). Inhibition of protein synthesis may be at translation or elongation of the polypeptide chain but the controls are not known; however, increased ion concentration and decreased ATP supply inhibit protein synthesis and could do so in stressed cells. In young tissues, stress can have less effect on protein synthesis than in old (Dungey & Davies, 1982) but this is not always observed (Pheloung & Barlow, 1981) due perhaps to differences in age of tissue and degree of stress. The capacity to degrade protein is smaller in young tissues so that stress has less effect on their protein content. ABA, which accumulates during stress, decreases protein synthesis at the translation stage and may be important in stressed tissues. Stress reduces the rate and amount of protein synthesis and changes the type of protein synthesised; gibberellic acid reverses the effects of ABA but not of stress (Dhindsa & Cleland, 1975). If structural proteins and enzymes are differentially synthesised under stress, it is to be expected that metabolism would change and allow regulation of plant response to stress but this potentially very important area of metabolism has not been studied. There is no evidence that stress changes the structure or kinetic constants of particular enzymes. Greater understanding of the effects of stress on protein synthesis is needed. With turn-over times of 90-220 hours for soluble proteins, faster for some regulatory enzymes (e.g. nitrate reductase) and slower for structural components (Dungey & Davies, 1982), loss of enzyme activity would quickly have major effects on metabolism.

Lipid Synthesis

Synthesis of complex lipids requires a supply of glyceraldehyde-3-phosphate and reductant (NADH) to give triglyceride and of acetyl CoA to produce fatty acids. Synthesis of phospholipids, which are important components of membranes, requires amino acids. Severe water stress has been reported to inhibit lipid synthesis and change the proportion of lipids in mature tissues (Chetal et al., 1981) but the mechanisms are poorly understood; control may operate at the level of enzyme synthesis or via the supply of NADH and ATP rather than on the reactions themselves. Inhibition of lipid synthesis is unlikely to be directly affected by changes in turgor or π and may be a consequence of alterations to other metabolic processes.

Pigment and Hormone Formation

Moderate stress inhibits synthesis of chlorophyll and carotenoids (see Fig. 4) which occurs after illumination of expanding etiolated leaves (Duysen & Freeman, 1975). Stress may inhibit δ-aminolevulinic acid (ALA) synthesis

Fig. 4. Chlorophyll and carotenoid synthesis, related to thylakoid formation and production of hormones. Large arrows show possible sites of control by stress.

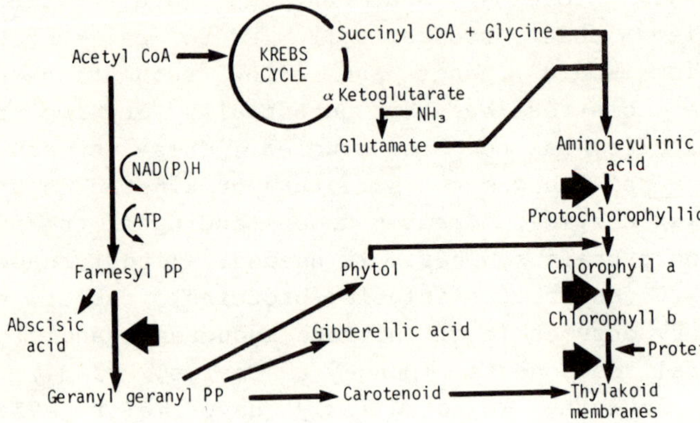

and its conversion to chlorophyll a. If chlorophyll b is derived from chlorophyll a then stress might decrease chlorophyll b more than chlorophyll a and increase the chlorophyll a/b ratio. Stress may also prevent synthesis of enzymes converting ALA to chlorophyll a and of proteins of the thylakoids, so that chlorophyll b does not accumulate and grana and thylakoid structure is altered (Duysen & Freeman, 1975). Photosystem 2 activity and processes in electron transport and water splitting may also be affected. However the origin of these fundamental changes in structure is obscure as they are the result of many complex metabolic events.

The phytohormones farnesol (in some species) and ABA accumulate in stressed leaves (Goodwin & Erwee, 1983) suggesting either that farnesyl pyrophosphate is more rapidly produced or its consumption is slowed with stress; the latter would decrease gibberellic acid (GA) and carotenoid biosynthesis (see Fig. 4) and cause an hormonal imbalance towards ABA and away from GA, slowing growth. However, there is little information on control points in terpenoid metabolism or the sensitivity of processes to stress, particularly in relation to leaf growth.

Cell Wall Synthesis

Water stress decreases glucose incorporation into cell walls less than it decreases expansion, (see Hsiao, 1973), suggesting that wall synthesis is relatively insensitive to mild stress (ψ above -1.0 to -1.5 MPa) but severe stress greatly inhibits incorporation. Cellulose and pectin synthesis from glucose-1-phosphate via GDP-glucose and UDP-galacturonate requires NADH and ATP, shortage of which could slow wall growth, but little is known of the control of these processes by P, π or specific ion effects in stressed, developing cells. It is not known if cessation of growth causes accumulation of products and feedback inhibition of enzyme activity or inhibition of enzyme synthesis or the function of microtubules at the cell surface, which may orientate cellulose in the wall.

Synthesis of Osmotically Compatible Solutes

Amino acids (proline particularly) and quaternary amino compounds (glycine betaine) are the principal organic, non-sugar osmoregulatory materials in stressed tissues. Proline is derived from glutamate in respiring tissues in reactions requiring NADH, and betaine from serine via intermediates of phospholipid synthesis (Fig. 5). Both processes may be linked to NADH concentration and electron transport in the mitochondria. Proline synthesis increases as cell expansion and protein synthesis are inhibited and may result from metabolic imbalance when nitrate or ammonia and reductant are available but normal synthetic processes are blocked. Betaine is a metabolic end product which accumulates in stressed tissues and may have a protective role (Hanson & Hitze, 1982).

When photosynthesising tissues are stressed, stomatal conductance falls, decreasing photosynthesis and increasing the RuBP oxygenase reaction relative to carboxylation. As a result relatively more carbon from the

Fig. 5. Scheme of synthesis of the osmoregulatory molecules betaine and proline, from storage materials.

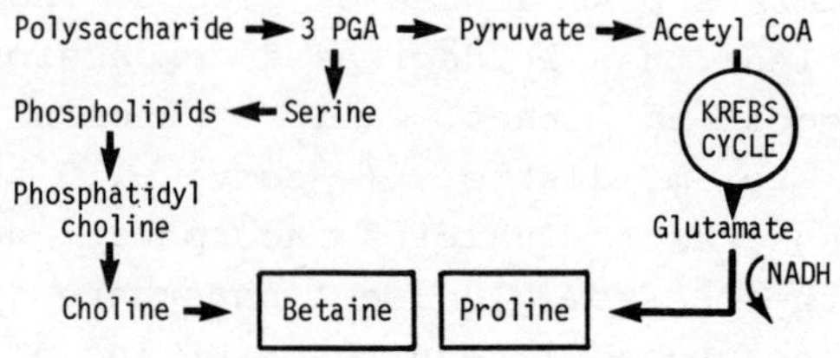

chloroplast enters the glycolate pathway in stressed than unstressed tissues (Lawlor, 1979) and glycine and serine accumulate. Serine metabolism is linked to that of choline and glycine-betaine (Fig. 5), so it is possible that in photosynthetically competent tissues this mechanism could provide carbon for synthesis of metabolically "compatible" solutes. Also ATP is synthesised in mitochondria as a consequence of glycolate metabolism and may help to maintain a large cytosolic adenylate energy charge (Atkinson, 1977). However this type of osmotic adjustment involves considerable changes in cell metabolism and may, in the short term, regulate the excess energy in the leaf rather than the osmotica. Therefore long-term adjustment to stress, particularly in non-photosynthetic cells, involves accumulation of ions, sugars, etc. and more subtle interaction between the "normal" control processes.

CONCLUSIONS

Growth of leaves and their final size are determined by the number of cells produced by cell division and the size to which cells expand. Expansion is regulated metabolically *via* cell wall binding characteristics and osmotic potential. Small water deficits in the environment decrease cell water content and turgor and slow cell expansion considerably, giving small cells with thick walls. However it is not established if the physical properties (*e.g.* bulk elasticity and yield threshold) of these cells are altered. Mild water stress has little effect on cell number or wall synthesis, although protein synthesis may be more sensitive. Leaves developed under stress have more dry matter per unit area than unstressed. Severe stress stops cell expansion, slows cell division and decreases wall growth, respiration and synthesis of proteins, lipids and pigments and inhibits membrane (*e.g.* thylakoid) formation. Synthetic processes, of which protein synthesis plays the central role in metabolism, may be regulated by "feedback" control of reactions or by increased concentration of effector molecules and ions, and by decreased ATP supply or enzyme concentrations. Metabolic products of synthetic pathways may accumulate as cell expansion is inhibited or may be synthesised *de novo* in reactions controlled by substrate availability and effector molecules. Regulation of metabolism under stress conditions is not well understood. Accumulated organic molecules, which are "compatible" with metabolism, and accumulated ions decrease osmotic potential, providing "osmotic adjustment" and increasing turgor, thus enabling growth and metabolism to be maintained under stress. Such changes in osmotic potential may be regulated by mechanisms which are triggered by turgor sensitive sensors. Upon relief of stress, rapid cell expansion occurs which may be due to large P and the ready availability of accumulated wall materials. Very small cell volume and turgor impair metabolism and damage leaf development as the capacity of metabolic controls to adjust to the imbalance is exceeded.

ACKNOWLEDGEMENTS

We thank Bill Day for stimulating discussions and Janet Why for help in preparation of the manuscript.

REFERENCES

Albersheim, P. (1976). The primary cell wall. In: Plant Biochemistry, Third Edition, eds. J. Bonner & J.E. Varner, pp. 225-74. London: Academic Press.

Atkinson, D.E. (1977). Cellular energy metabolism and its regulation. New York: Academic Press.

Barlow, E.W.R., Munns, R.E. & Brady, C.J. (1980). Drought responses of apical meristems. In: Adaptation of Plants to Water and High Temperature Stress, eds, N.C. Turner & P.J. Kramer, pp. 191-205. New York: Wiley - Interscience.

Boyer, J.S. & Younis, H.M. (1983). Molecular aspects of photosynthesis at low leaf water potentials. In: Effects of stress on photosynthesis, eds. R. Marcelle, H. Clysters & M. van Pouche, pp. 29-33. The Hague: Martinus Nijhoff/Dr. W. Junk, Publishers.

Chetal, S., Wagle, D.S. & Nainawatee (1981). Glycolipid changes in wheat and barley chloroplast under water stress. Plant Sci. Letts., 20, 225-30.

Cleland, R.E. (1977). The control of cell enlargement. In: Integration of Activity in the Higher Plant, ed. D.H. Jennings. Symposia of the Society for Experimental Biology, No. 31, pp. 101-15. Cambridge: Cambridge University Press.

Cutler, J.M., Rains, D.W. & Loomis, R.S. (1977). The importance of cell size in the water relations of plants. Physiol. Plant., 40, 225-60.

Cutler, J.M., Shahan, K.W. & Steponkus, P.L. (1980). Influence of water deficits and osmotic adjustment on leaf elongation in rice. Crop Sci., 20, 314-18.

Delane, R., Greenway, H., Munns, R. & Gibbs, J. (1982). Ion concentration and carbohydrate status of the elongating tissue of Hordeum vulgare growing at high external NaCl. J. Exp. Bot., 33, 557-73.

Dhindsa, R.S. & Cleland, R.E. (1975). Water stress and protein synthesis. Plant Physiol., 55, 782-85.

Dungey, N.O. & Davis, D.D. (1982). Protein turnover in the attached leaves of non-stressed and stressed barley seedlings. Planta, 154, 435-40.

Duysen, M.E. & Freeman, T.P. (1975). Partial restoration of the high rate of plastid pigment development and the ultrastructure of plastids in detached water-stressed wheat leaves. Plant Physiol., 55, 768-73.

Goodwin, P.B. & Erwee, M.G. (1983). Hormonal influences on leaf growth In: The Growth and Functioning of Leaves, eds. J.E. Dale & F.L. Milthorpe, pp. 207-32. Cambridge: Cambridge University Press.

Hanson, A.D. & Hitz, W.D. (1982). Metabolic responses of mesophytes to plant water deficits. Ann. Rev. Plant Physiol., 33, 163-203.

Henckel, P.A. (1964). Physiology of plants under drought. Ann. Rev. Plant Physiol., 15, 363-86.

Hsiao, T.C. (1973). Plant responses to water stress. Ann. Rev. Plant Physiol., 24, 519-70.

Larkum, A.W.D. & Wyn Jones, R.G. (1979). Carbon dioxide fixation by chloroplasts isolated in glycinebetaine. Planta, 145, 393-94.

Lawlor, D.W. (1972). Growth and water use of Lolium perenne. II. Plant growth. J. Appl. Ecol., 9, 99-105.

Lawlor, D.W. (1979). Effects of water and heat stress on carbon metabolism of plants with C_3 and C_4 photosynthesis. In: Stress Physiology in Crop Plants, eds. H. Mussell & R.C. Staples, pp. 303-326. New York: John Wiley & Sons.

Lawlor, D.W. (1983). Integration of biochemical processes in the physiology of water stressed plants. In: Effects of Stress on Photosynthesis, eds. R. Marcelle, H. Clijsters & M. Van Poucke, pp. 35-44. The Hague: Martinus Nijhoff/Dr W. Junk, Publishers.

Legg, B.J., Day, W., Lawlor, D.W. & Parkinson, K.J. (1979). The effects of drought on barley growth: models and measurements showing the relative importance of leaf area and photosynthesis. J. Agric. Sci., Camb., 92, 703-16.

Lockhart, J.A. (1965). Cell extension. In: Plant Biochemistry, eds. J. Bonner & J.E. Varner, pp. 826-49. London: Academic Press.

Lyndon, R.F. (1983). The mechanism of leaf initiation. In: The Growth and Functioning of Leaves, eds. J.E. Dale & F.L. Milthorpe, pp. 3-24. Cambridge: Cambridge University Press.

Munns, R. & Weir, R. (1981). Contribution of sugars to osmotic adjustment in elongating and expanding zones of wheat leaves during moderate water deficits at two light levels. Aust. J. Plant Physiol., 8, 93-105.

Nobel, P.S. (1974). Introduction to Biophysical Plant Physiology. San Francisco: W.H. Freeman.

Pheloung, P. & Barlow, E.W.R. (1981). Respiration and carbohydrate accumulation in the water-stressed wheat apex. J. Exp. Bot., 32, 921-31.

Preiss, J. & Kosuge, T. (1976). Regulation of enzyme activity in metabolic pathways. In: Plant Biochemistry, Third Edition, eds. J. Bonner & J.E. Varner, pp. 277-336. London: Academic Press.

Robards, A.W. (1970). Electron Microscopy and Plant Ultrastructure. London: McGraw-Hill.

Terry, N., Waldron, L.J. & Taylor, S.E. (1983). Environmental influences on leaf expansion. In: The Growth and Functioning of Leaves, eds. J.E. Dale & F.L. Milthorpe, pp. 179-205. Cambridge: Cambridge University Press.

Turner, N.C. & Jones, M.M. (1980). Turgor maintenance by osmotic adjustment: a review and evaluation. In: Adaptation of Plants to Water and High Temperature Stress, eds. N.C. Turner & P.J. Kramer, pp. 87-103. New York: John Wiley & Sons.

Tyree, M.T. & Jarvis, P.G. (1982). Water in tissues and cells. In: Physiological Plant Ecology. II. Water Relations and Carbon Assimilation, eds. O.L. Lange, P.S. Nobel, C.B. Osmond & H. Ziegler, pp. 35-77. New York: Springer-Verlag.

Wyn Jones, R.G. & Gorham, J. (1983). Osmoregulation. In: Encyclopedia of Plant Physiology. N.S. Vol. 12. Physiological Plant Ecology III, eds. O.L. Lange, P.S. Nobel, C.B. Osmond & H. Ziegler, pp. 35-58. Berlin: Springer-Verlag.

Zimmermann, V. (1977). Cell turgor pressure regulation and turgor pressure-mediated transport processes. In: Integration of Activity in the Higher Plant, ed. D.H. Jennings. Symposia of the Society for Experimental Biology, No. 31, pp. 117-54. Cambridge: Cambridge University Press.

AIR POLLUTION AND LEAF GROWTH

Mary E. Whitmore, W.J. Davies and T.A. Mansfield
Department of Biological Sciences,
University of Lancaster, Lancaster, LA1 4YQ,
United Kingdom.

INTRODUCTION

The air of industrialised countries contains an array of pollutants which are potentially toxic to plants, and the most widespread of these are ozone, sulphur dioxide and oxides of nitrogen. In parts of the United States the summer-time concentrations of ozone are sufficient to cause extensive damage to the leaves of particularly sensitive species and there has been considerable interest in the effects of this pollutant on plants. In western Europe, however, episodes of ozone pollution are less frequent and much greater attention has been directed to sulphur dioxide and, more recently, to mixtures of sulphur dioxide and nitrogen dioxide. We will limit our discussion to the effects of these three pollutants on leaf growth in the hope that we can characterise similarities and differences which may be representative of pollution injury in general.

During the last twenty years there has been a gradual change in emphasis from the study of acute leaf injury arising from exposure to high concentrations of pollutants (>1 ppm) to the more subtle effects of exposure to low concentrations for long periods. This transition has occurred hand-in-hand with the development of better exposure techniques and an improved characterisation of the polluted atmosphere.

Sulphur dioxide in the atmosphere arises from the burning of fossil fuels, all of which contain some sulphur compounds, and from the smelting of sulphur ores. During the past three decades the pattern of dispersion of sulphur dioxide has altered because of the building of much taller industrial stacks. Concentrations close to sources have tended to fall and sulphur dioxide is now much more widely dispersed. The burning of fossil fuels is also a major source of oxides of nitrogen but, unlike sulphur dioxide, oxides of nitrogen are formed from atmospheric nitrogen which is oxidised in the heat of a flame. Nitric oxide is formed initially and is then slowly oxidised in the atmosphere to nitrogen dioxide. It has been estimated that in north-west Europe up to half the oxides of nitrogen is emitted from car exhausts and the remainder from the industrial and domestic burning of fuels (Soderland, 1977).

There is still a dearth of information regarding the concentrations of sulphur dioxide and nitrogen dioxide that occur over agricultural land but Fowler & Cape (1982) have used the existing monitoring data to estimate distribution zones in western Europe. They suggest that these two pollutants have similar distributions with highest concentrations in urban areas, and even in some rural land adjacent to towns annual means of greater than 0.04 ppm of each gas may occur. Annual mean concentrations do not, however, fully describe the distribution, for both sulphur dioxide and nitrogen dioxide show a marked annual cycle with higher concentrations in winter months, reflecting the increased use of fuels at this time of year.

In contrast to sulphur dioxide and nitrogen dioxide, elevated ozone concentrations predominantly occur in the summer months. Relatively high concentrations occur naturally in the upper atmosphere and during vertical mixing some is brought down to ground level, so that even "clean" air may have up to 0.04 ppm ozone. Higher levels are formed by the action of sunlight on nitrogen oxides and hydrocarbons emitted from vehicle exhausts, so that many locations in North America and Europe with heavy motor traffic may experience concentrations of up to 0.2 ppm ozone during the summer (Vukovich et al., 1977; Ball, 1976). Because ozone can be transported large distances from its origins in cities (Cox et al., 1975) elevated concentrations may even occur in rural areas (Harrison & Holman, 1979). We can conclude that some agricultural areas in industrialised countries may regularly experience elevated concentrations of sulphur dioxide, nitrogen dioxide and/or ozone of up to 0.07 ppm, and on a daily or hourly basis concentrations may occasionally rise even higher.

There are now several reports of the effects on growth of prolonged exposure to realistic concentrations of ozone and of sulphur dioxide applied separately. In a few cases the detailed responses of whole plants have been followed at intervals and these studies have revealed interesting effects on leaf growth and morphology. It appears that in some cases a pollutant-induced reduction in net assimilation rate is countered by an increase in the relative size of the assimilatory area. This is brought about by the maintenance of leaf growth, often at the expense of the roots, and in some cases by the production of leaves of lower density. There have been relatively few studies of the responses to oxides of nitrogen, and generally plants seem to be less sensitive to oxides of nitrogen than to either ozone or sulphur dioxide. The available information suggests that similar responses do not occur in nitrogen dioxide, but that its presence in a mixture of sulphur dioxide and nitrogen dioxide may accentuate the responses that are characteristic of sulphur dioxide. In this chapter, examples are chosen to illustrate our present understanding of such responses, and the possible consequences for the plant when exposed to other environmental stresses are then discussed.

INCREASED INVESTMENT IN LEAF AREA BY POLLUTED PLANTS

The magnitude of the effects of sulphur dioxide or ozone on growth depends not only on the concentration and the species and variety, but also on the environmental conditions during exposure and the physiological stage of development of the plant. There has therefore been considerable difficulty in establishing a threshold concentration below which plant growth is unaffected, even when a single variety which has been extensively investigated (Bell, 1982). A further complicating factor in cases where only the shoots of treated and untreated plants have been compared is the existence of certain growth compensation mechanisms by which polluted plants invest in leaf area to a greater extent than controls.

The leaves of plants exposed to nitrogen dioxide often appear greener than those of controls, but leaves exposed to ozone or to sulphur dioxide usually appear chlorotic and tend to senesce more rapidly. Davies (1980) illustrated this increased senescence in leaves of *Phleum pratense* exposed continuously to 0.12 ppm sulphur dioxide under controlled environmental conditions conducive to slow growth. Measurements of green leaf length were made at weekly intervals during a six-week exposure (Fig. 1). The first and second leaves expanded normally in sulphur dioxide and attained the same maximum length as controls, but they started to senesce earlier and more rapidly. Growth of the subsequent leaves was also affected; the development of the third, to sixth leaves was delayed in sulphur dioxide so that the control plants had six leaves at the end of the experiment but the sulphur dioxide-treated plants had only five.

When the pollutant stress is less severe, increased senescence of older leaves is sometimes accompanied by a stimulation of leaf production. This was observed by Bennett & Oshima (1976) when *Daucus carota* was exposed intermittently to 0.19 or 0.25 ppm ozone during a 108-day growth period (Table 1). Although exposure to 0.25 ppm ozone caused chlorosis on 28% of the leaves, more leaves were produced and the length of the largest leaf increased by 15%. This did, however, involve a cost to the plant which in this case was a considerable inhibition of growth of the hypocotyl. Clearly there was an alteration in sink priorities within the plant which favoured new leaves. This type of experiment does not, however, give an insight into the causes of these changes and only by following growth at regular intervals and using growth analysis have the

Fig. 1. Changes in green leaf length of Phleum pratense when exposed to clean air (——) or 0.12 ppm sulphur dioxide (----) for six weeks. The asterisks denote the leaves of polluted plants. (From Davies, 1980).

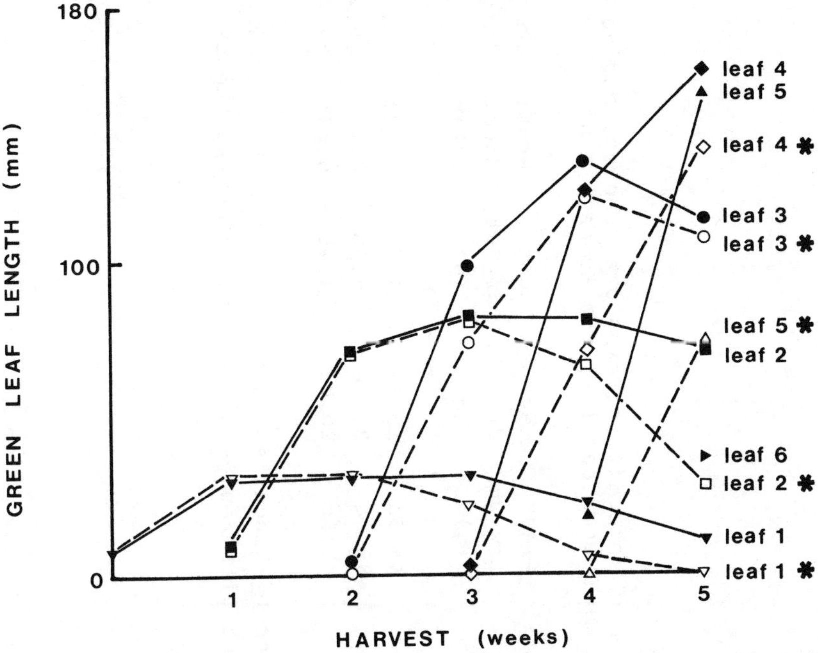

dynamics of responses been elucidated.

When the growth of Petroselinum crispum was followed during exposure to ozone, Oshima et al. (1978) noted that although relative growth rate (RGR) was initially depressed by ozone, it recovered as the exposure continued. In a subsequent experiment Oshima et al. (1979) calculated net assimilation rates (NAR) of Gossypium hirsutum exposed to ozone and observed that although the NAR was consistently depressed by ozone, an increase in the leaf area ratio (LAR) of polluted plants allowed the RGR to recover. Clear confirmation of these compensatory responses

Table 1. Effects of intermittent exposure to ozone during a 108-day growth period on Daucus carota. From Bennett & Oshima (1976).

Ozone (ppm)	Total no. leaves	% Chlorotic leaves	Plant dry wt. (g)	Leaf dry wt. (g)	Root dry wt. (g)	Root: Shoot ratio	Length of largest leaf (cm)
CONTROL	34.1a*	2a	44.5a	19.6a	24.9a	1.23a	45.7a
0.19	39.2a	14b	39.1a	22.2a	16.9b	0.73b	50.2ab
0.25	39.8a	28c	32.6b	20.0a	13.4b	0.60b	52.4b

* Mean separation in columns by Duncan's multiple range test. Different subscripts denote significant differences between the means at the 5% level of probability.

has been provided by Walmsley <u>et al</u>. (1980) who followed the development of <u>Raphanus sativus</u> in clean air or in air containing 0.17 ppm ozone in the laboratory, and applied functional growth analysis techniques to the results. Fig. 2 shows the change in leaf and hypocotyl dry weights with time. Although leaf dry weight was initially lower in ozone, there was little difference between the control and polluted leaf weights after 30 days. Hypocotyl dry weight, however, was considerably lower in the ozone-treated plants throughout the exposure. The calculation of whole plant RGR, NAR and LAR revealed that ozone depressed NAR throughout the middle part of the exposure and RGR was initially

Fig. 2. Growth of the leaf and hypocotyl of Raphanus sativus in clean air (●) or in 0.17 ppm ozone (○). Each point is the mean of six replicates. (From Walmsley et al., 1980).

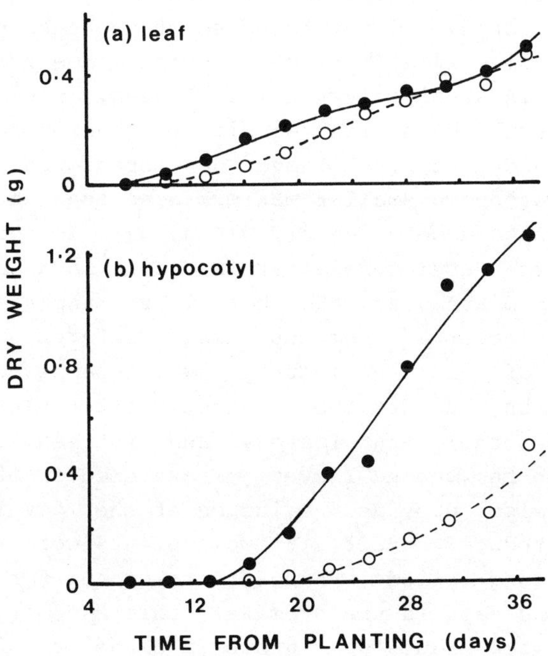

depressed (Fig. 3). However, an increase in the relative size of the assimilatory area more than compensated for the reduced efficiency of the leaves per unit area and during the latter part of the exposure, RGR in ozone slightly exceeded that of the control.

In these examples, increases in LAR were brought about by the greater investment in leaf dry weight by the polluted plants, but in some cases higher specific leaf area (SLA) has also contributed to increases in LAR. This was observed by Jones & Mansfield (1982) when Phleum pratense was exposed to 0.12 ppm sulphur dioxide in controlled conditions. Comparable responses to those seen in ozone occurred in response to sulphur dioxide (Fig. 4) but the higher LAR in sulphur dioxide towards the end of the exposure was caused not only by a relatively higher production of leaf dry matter, at the expense of root growth, but also by an increase in SLA.

The study of the response of Raphanus sativus to ozone by Walmsley et al. (1980) provided some evidence for differences in sensitivity between leaves, with successive leaves apparently being less sensitive. They noted that the cotyledons and the first leaves developed more slowly in ozone and reached a smaller maximum area than controls, and they then senesced more rapidly (Fig. 5). In contrast, the second leaves developed faster in ozone and attained a greater maximum area, and the third leaves appeared earlier. They also observed corresponding differences in the development of visible injury, which appeared almost immediately on the cotyledons but only after nine days from emergence on the first leaves, and fifteen days after emergence on the second leaves. Measurements of stomatal resistance also provided evidence of the development of leaves with reduced sensitivity when plants were exposed to ozone for long periods. Exposure to ozone for 48 hours caused an increase in the stomatal resistance of the second leaves, but when plants were grown in ozone for 30 days the stomatal resistance of these leaves did not differ from controls. Clearly, there is evidence of several adaptive

Fig. 3. Mean growth parameters for Raphanus sativus in clean air (●) or in 0.17 ppm ozone (○). (a) Relative growth rate (b) Net assimilation rate; (c) Leaf area ratio. The values and 95% confidence limits are those of curves fitted to the original data. (From Walmsley et al., 1980).

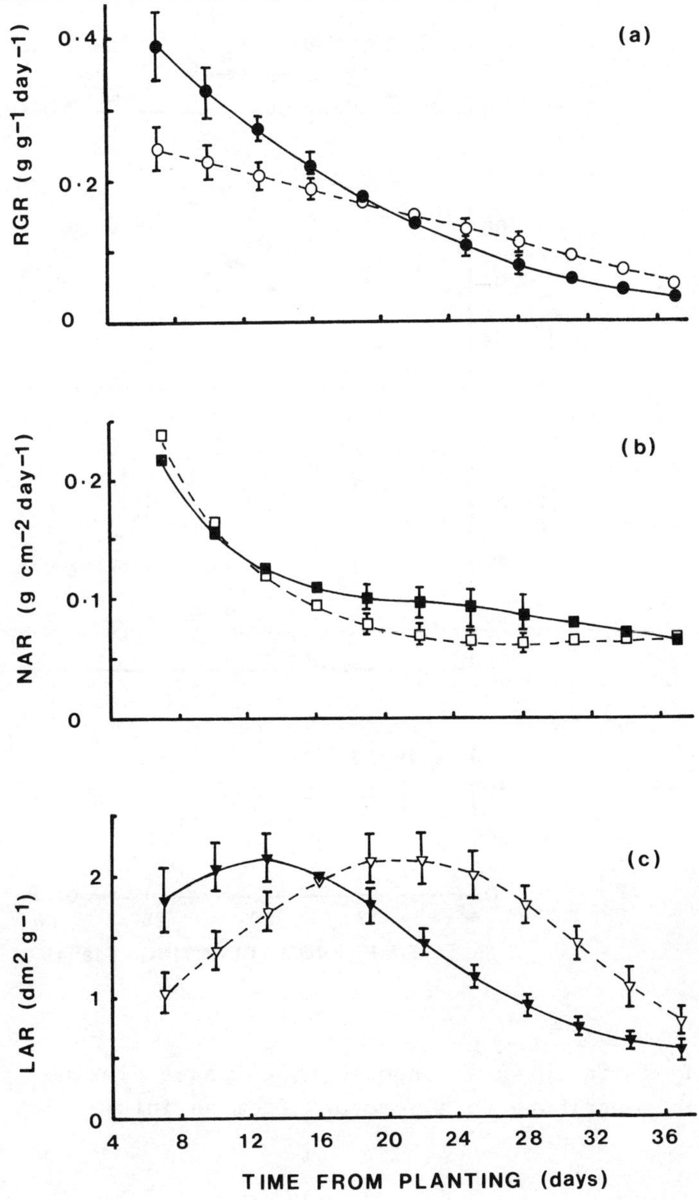

Fig. 4. Development of area of individual leaf pairs of Raphanus sativus in clean air (●) or in 0.17 ppm ozone (○). Each point is the mean of six replicate plants. (From Walmsley et al., 1980).

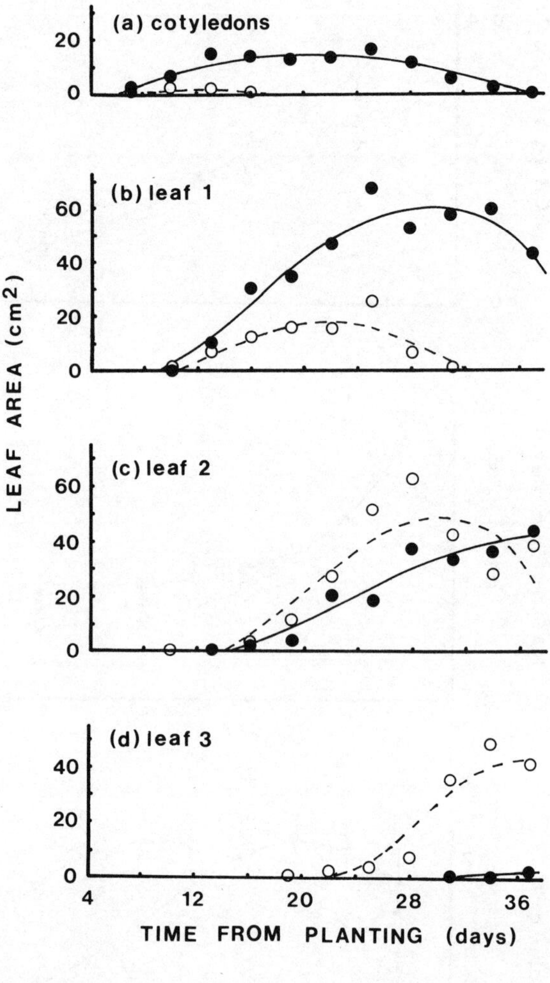

changes in plants exposed to sulphur dioxide or ozone which contribute to a recovery from an initial depression in RGR.

*Fig. 5. Mean growth parameters for Phleum pratense in clean air (●) or 0.12 ppm sulphur dioxide (○) in the laboratory. (a) Relative growth rate; (b) Net assimilation rate; (c) Leaf area ratio; (d) Specific leaf area. Each point is the mean value for eight replicate plants. Significance of sulphur dioxide effect is shown with asterisks; *P<0.05. (From Jones & Mansfield, 1982).*

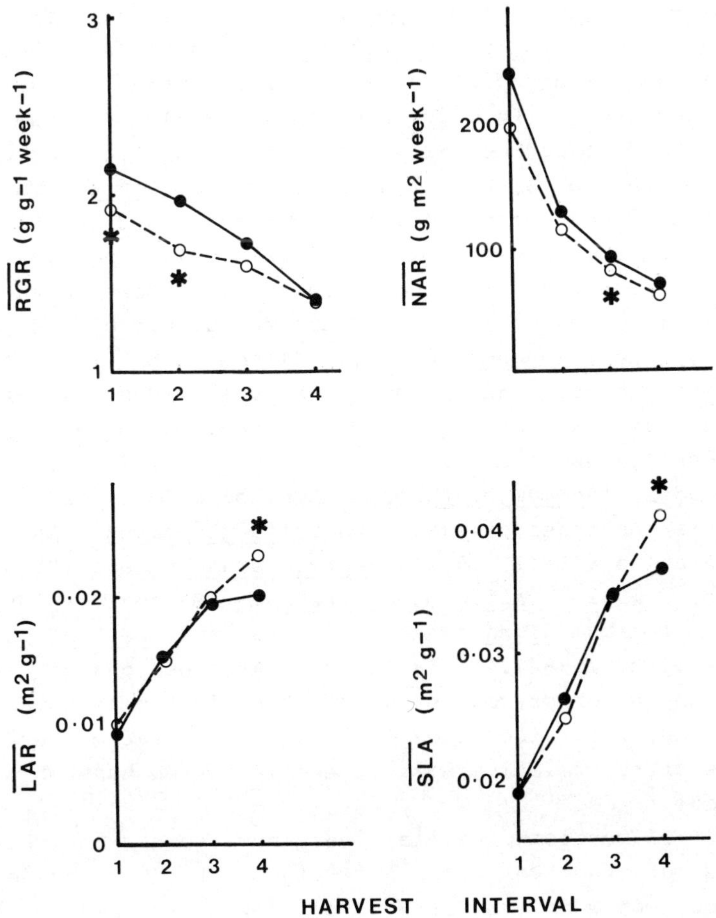

There have been few detailed studies of this kind and it is therefore difficult to assess how widespread such changes are. However, many experiments where plant growth has been measured at the end of an exposure have shown that the leaves of plants exposed to ozone or sulphur dioxide are often the least affected organs in terms of dry weight. For example, smaller reductions in shoot than in root growth have also been observed after exposure to ozone in _Phaseolus vulgaris_ (Adedipe _et al_., 1972), _Trifolium incarnatum_ and _Lolium multiflorum_ (Bennett & Runeckles, 1977), _Festuca arundinaceae_ (Flagler & Youngner, 1982) and _Pinus ponderosa_ (Parmeter & Miller, 1968). It has been demonstrated experimentally that this is not a direct effect on the roots, for when the leaves of _Glycine max_ were protected so that ozone could only reach the plant via the soil, the roots were unaffected (Blum & Tingey, 1977). Smaller reductions in shoot than in root growth have also been widely observed after exposure to sulphur dioxide, for example, in _Raphanus sativus_ (Tingey _et al_., 1971), _Nicotiana tabacum_ and _Cucumis sativus_ (Mejstrik, 1980), _Glycine max_ (Reinert & Weber, 1980), _Lolium perenne_ (Ashenden & Mansfield, 1977; Bell _et al_., 1979; Ayazloo _et al_., 1980), _Phleum bertolonii_ (Ayazloo & Bell, 1981), _Poa pratensis_ (Ashenden, 1979), _Dactylis glomerata_ (Ashenden, 1979; Ayazloo & Bell, 1981), _Festuca arundinaceae_ (Flagler & Youngner, 1982), _Ailanthus altissima_ (Marshall & Furnier, 1981), _Populus nigra_, _Alnus incana_ and _Betula pendula_ (Freer-Smith, 1984). It is likely that these changes in the partitioning of dry matter within the plant will have been associated with increases in leaf area ratio but this measure of the relative size of the leaf area has not always been made.

As mentioned earlier there have been relatively few studies of the responses of plants to the oxides of nitrogen, but where the effects on growth have been examined the responses do not appear to be comparable with those observed in ozone or sulphur dioxide. Whitmore & Mansfield (1983) made a direct comparison of the effects of sulphur dioxide and nitrogen dioxide on the growth of *Poa pratensis* exposed in unheated glasshouses with natural illumination. In terms of effects on total plant dry weight, *P. pratensis* appeared to be equally sensitive to sulphur dioxide and nitrogen dioxide applied individually; the reductions in plant size were evident at precisely the same time and were almost identical in magnitude (Fig. 6). However, whilst sulphur dioxide caused an increase in shoot:root weight ratio, there were no significant alterations in dry matter partitioning in plants exposed to nitrogen dioxide. When the two gases were supplied together the effects on growth were more severe than if either gas was applied singly, and the alteration in dry matter partitioning was more marked. These changes in partitioning contributed to higher leaf area ratios in sulphur dioxide and in sulphur dioxide plus nitrogen dioxide but there was also evidence that these plants had higher specific leaf areas.

Increases in SLA have also been observed in *Lolium perenne* (Bell *et al*., 1979) and *Populus deltoides* (Jensen, 1981) after exposure to sulphur dioxide, but such changes in the plant as a whole are complex because of differences in sensitivity between leaves, and because older leaves may have senesced so that leaves at different stages of development are being compared. We have recently examined how increases in SLA of the most recently expanded leaf on the main tiller of *Phleum pratense* exposed to sulphur dioxide plus nitrogen dioxide are brought about. Table 2 shows the leaf areas, fresh and dry weights and the SLA of this leaf when plants were exposed to 0.09 ppm sulphur dioxide and 0.1 ppm nitrogen dioxide for nineteen days. The SLA of the polluted leaves was 122% of that of the controls, and was caused by the polluted leaves having a smaller dry weight spread over a larger area. The larger area was a result of the increase in leaf length as there was no effect on leaf width. Although the dry weight of the polluted leaves was smaller, their fresh weights were almost identical to the controls. Consequently, the polluted leaves

Fig. 6. Growth of Poa pratensis in clean air (●) or in weekly mean concentrations of 0.062 ppm nitrogen dioxide (□), 0.062 ppm sulphur dioxide (△) or 0.062 ppm sulphur dioxide plus 0.062 ppm nitrogen dioxide (○). Exposures were done in glasshouses during the winter and spring of 1979/80. (a) Plant dry weight; (b) shoot:root ratio. Bars represent least significant differences, p = 0.05. (From Whitmore & Mansfield, 1983).

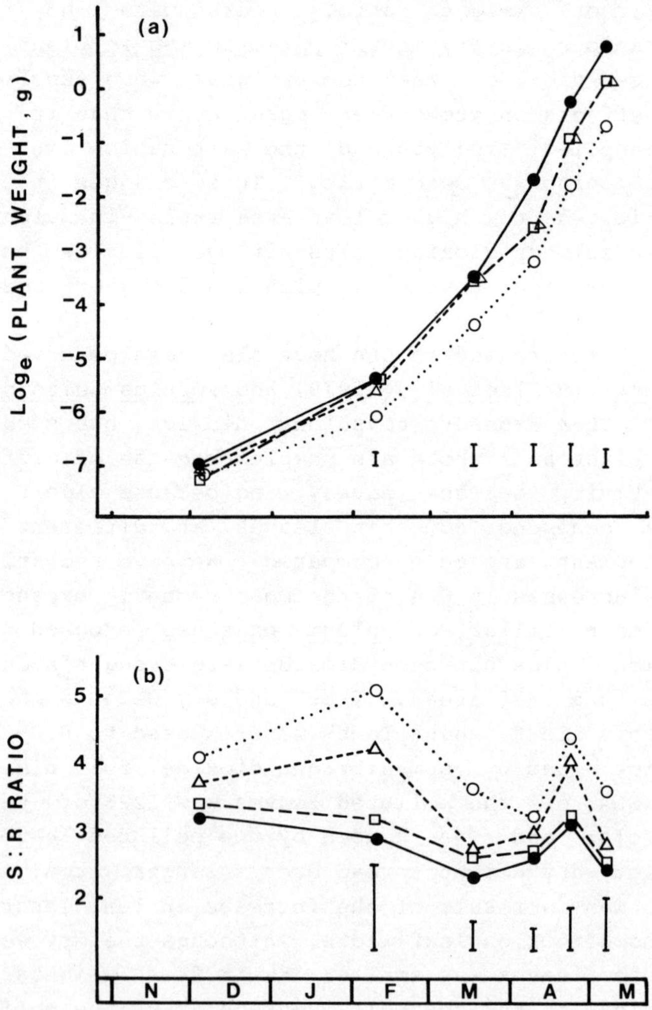

also had a higher moisture content. Histological observations suggested that the mesophyll cells of the polluted leaves were larger and less tightly packed than controls (Plate 1). Measurements of the water relations of these leaves revealed that osmotic potentials were less negative in polluted leaves with the result that turgors were consistently slightly lower than those of plants in clean air (Table 3).

MECHANISMS OF GROWTH COMPENSATION RESPONSES

The application of growth analysis techniques has revealed that ozone, sulphur dioxide and nitrogen dioxide reduce the productive efficiency of the green leaves on a unit area basis. This may be due to direct or indirect action on photosynthetic processes or to an increased usage of photosynthates for the repair of injury and detoxification of the pollutant derivatives. Effects of these pollutants on physiological processes and plant metabolism have recently been extensively reviewed (Koziol & Whatley, 1984) and there is evidence that low concentrations of all three pollutants may inhibit net photosynthesis (Bull & Mansfield, 1974; Black & Unsworth, 1979; Black *et al.*, 1982). The subsequent reduction in available assimilates could lead directly to an alteration in dry matter distribution, for when environmental conditions cause assimilate deficiency, growth of the shoot appears to have priority over that of the root (Wardlaw, 1968). Similarly, a reduction in assimilate availability or composition caused by infection by foliar pathogens, such as the powdery mildews and rusts, is generally considered to be the cause of the greater inhibition of root than of shoot growth in infected plants (Siddiqui & Manners, 1971; Baenziger *et al.*, 1979). It appears, therefore, that an increased investment in leaf growth can be invoked by several agents other than air pollutants in response to an interference with assimilatory activity.

In contrast, when plant growth has been inhibited by nitrogen dioxide, effects on the various plants part have

Table 2. Effects of exposure to 0.09 ppm sulphur dioxide and 0.1 ppm nitrogen dioxide in combination for 19 days on the most recently fully expanded leaf on the main tillers of Phleum pratense. Mean values with standard errors are given.

	Area (mm^2)	Length (mm)	Width (mm)	Fresh Weight (mg)	Dry Weight (mg)	Moisture Content (%)	SLA ($mm^2 mg^{-1}$)
CONTROL	2115+116	345±11	7.0±0.3	311±18	56±4	469±16	39.2±1.7
POLLUTED	2344+ 91	381±11	7.1±0.2	312±17	50±3	529±14	48.0±1.7
% OF CONTROL	111%	110%	101%	100%	89%	113%	122%
SIGNIFICANCE	NS	*	NS	NS	NS	**	***

NS = not significant,
* = significant at P <0.05,
** = significant at P <0.01,
*** = significant at P <0.001

Plate 1. Cross-sections of leaves of *Phleum pratense* sampled mid-way between the leaf tip and the ligule. (a) Leaf grown in clean air x 123; (b) leaf grown in 0.09 ppm sulphur dioxide plus 0.1 ppm nitrogen dioxide 123 (c) leaf grown in 0.09 ppm sulphur dioxide plus 0.1 ppm nitrogen dioxide x 309. Nute the large air spaces in the polluted leaves.

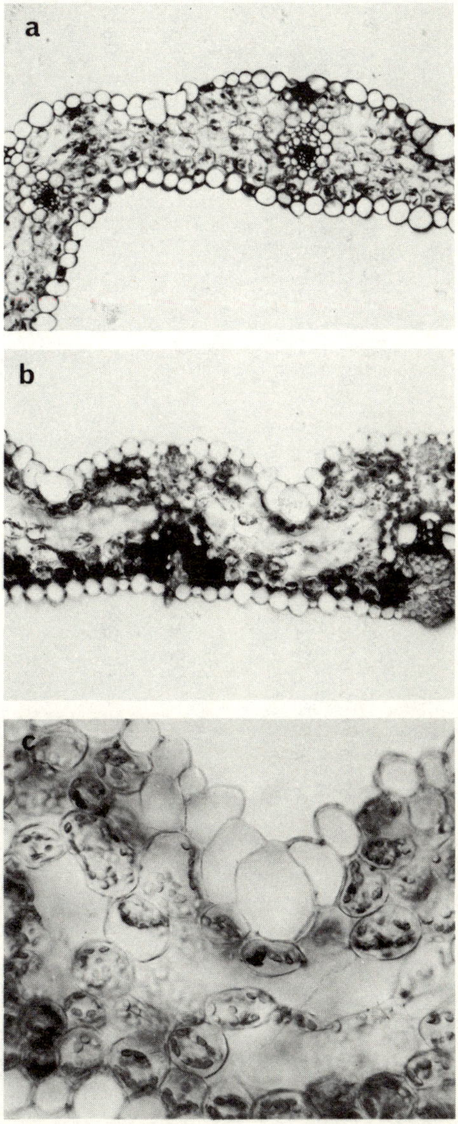

Table 3. Effects of exposure to 0.09 ppm sulphur dioxide and 0.1 ppm nitrogen dioxide in combination for 25 days on the water relations of the most recently expanded leaf on the main tillers of Phleum pratense. ψ, water potential; π, osmotic potential; P, turgor pressure. Mean values with standard errors are given.

	ψ (MPa)	π (MPa)	P (MPa)
CONTROL	−0.39±0.03	−1.05±0.03	0.66±0.05
POLLUTED	−0.39±0.02	−0.96±0.03	0.57±0.03

been of similar magnitude (Capron & Mansfield, 1977; Ashenden, 1979; Whitmore & Mansfield, 1983). It seems that growth compensation responses similar to those induced by sulphur dioxide and ozone are not brought into play despite the fact that nitrogen dioxide is known to inhibit net photosynthesis.

The SLA is a very plastic parameter which is prone to ontogenetic drift and sensitive to environmental change. Shade leaves often have higher SLA than sun leaves, and in monocotyledons they may also be longer. These changes are associated with a smaller palisade mesophyll layer and differences in photosynthetic properties which allow the plant to adapt to shade conditions. The increases in SLA and leaf length observed in sulphur dioxide appear to be associated with larger, less tightly packed mesophyll cells which have reduced turgor. One effect of the pollutant may be to loosen the mesophyll cell walls, leading to larger cells with less turgor. Presumably, turgor in these large cells remains low at least partly because of a limited supply of photosynthate. It is well known that acid-induced loosening of cell walls is a prerequisite for cell growth (see Chapter 11), and it is interesting to speculate, therefore, that an acid gas might influence cell wall properties to increase cell size and, ultimately, leaf size. Both sulphur dioxide and nitrogen dioxide dissociate in solution to form weak acids and might, therefore, be expected to cause acid-induced loosening of cell walls. However, it is possible that reduced amines may be formed in some circumstances in the chloroplasts of leaves exposed to nitrogen dioxide, as a result of various processes, and these may neutralise acidity at the cell wall. Zeevaart (1976) noted that the acid-induced necrotic leaf injury arising from exposure to nitrogen dioxide was markedly reduced when plants were exposed to a mixture of ammonia and nitrogen dioxide. In contrast, the reduced sulphur compounds resulting from incorporation of sulphur from sulphur dioxide may be emitted from leaves in measurable quantities (Hallgren et al., 1982). These compounds have no neutralising properties and thus pH at the cell walls may decrease.

POSSIBLE CONSEQUENCES FOR THE POLLUTED PLANT

Changes in the distribution of assimilate within the plant and in leaf morphology may allow the relative growth rate to recover and even to exceed that of the controls, but these changes may also have undesirable consequences. For example, in conditions of water shortage plants with a higher shoot:root ratio may be less able to maintain turgor: they not only have a relatively smaller root system with which to absorb water, but they also have a relatively larger leaf area from which to transpire. The reduction in size of the root system may also affect the ability of the roots to maintain nutrient supplies and could delay recovery after defoliation. Under these conditions the polluted plants may therefore be at a considerable disadvantage.

The maintenance of leaf growth by polluted plants may occur at the expense of an organ which is economically important, such as the hypocotyl of *Raphanus sativus* and *Daucus carota* exposed to ozone (Bennett & Oshima, 1976; Walmsley *et al*., 1980). This may not be confined to root crops, for boll production in *Gossypium hirsutum* and fruit production in *Capsicum annum* were reduced as a result of altered partitioning when plants were exposed to ozone (Oshima *et al*., 1979; Bennett *et al*., 1979), and bud formation was delayed when *Helianthus annuus* was exposed to sulphur dioxide (Shimizu *et al*., 1980).

An increase in SLA may also influence plant survival under sub-optimal environmental conditions, as a combination of large cells and low solute concentrations may not be compatible with the process of frost-hardening and might lead to the increases in frost sensitivity which have recently been reported (Davison & Bailey, 1982; Baker *et al*., 1982). Furthermore, a higher SLA could result in a greater uptake of pollutants per unit dry weight, thus counteracting the advantage of having an increased photosynthetic area.

Although these side-effects of the growth compensation responses are largely hypothetical and have not yet been investigated in detail, there is evidence from other areas of physiological ecology to lend support to these suggestions. For example, an altered dry matter partitioning in response to infection by foliar pathogens is considered to contribute to a lowering of crop yield under conditions of water shortage.

CONCLUSIONS

We have identified the existence of certain growth compensation mechanisms which serve to increase the relative leaf area of plants exposed to ozone and sulphur dioxide, and allow inhibitory effects on the efficiency of the leaves to be overcome. These responses may not, however, be representative of all types of pollution injury because there is evidence that plants do not respond in this way to nitrogen dioxide. There is also some evidence that polluted plants can produce new leaves that are less sensitive to pollutant injury. These changes could eventually lead to a recovery of growth under optimal conditions, but they may well have less desirable consequences under less favourable conditions. The investigation of these possible consequences represents a new area of pollution research.

ACKNOWLEDGEMENT

We are grateful to the Natural Environment Research Council for their support of the experimental work of M.E. Whitmore.

REFERENCES

Adepipe, N.O., Hofstra, G. & Ormrod, D.P. (1972). Effects of sulphur nutrition on phytotoxicity and growth responses of bean plants to ozone. Can. J. Bot., 50, 1789-93.

Ashenden, T.W. (1979). The effects of long-term exposures to SO_2 and NO_2 pollution on the growth of *Dactylis glomerata* L. and *Poa pratensis* L. Environ. Pollution (Series A), 18, 249-58.

Ashenden, T.W. & Mansfield, T.A. (1977). Influence of windspeed on the sensitivity of ryegrass to SO_2. J. Exp. Bot., 28, 729-35.

Ayazloo, M. & Bell, J.N.B. (1981). Studies on the tolerance to sulphur dioxide of grass populations in polluted areas. I. Identification of tolerant populations. New Phytol., 88, 203-22.

Ayazloo, M., Bell, J.N.B. & Garsed, S.G. (1980). Modification of chronic sulphur dioxide injury to *Lolium perenne* L. by different sulphur and nitrogen nutrient treatments. Environ. Pollution (Series A), 22, 295-307.

Baenziger, P.S., Kilpatrick, R.A. & Moseman, J.G. (1979). Reduced root and shoot growth in nutrient caused by *Erysiphe graminis tritici* in related wheats grown in nutrient solution culture. Can. J. Bot., 57, 1345-8.

Baker, C.K., Unsworth, M.H. & Greenwood, P. (1982). Leaf injury on wheat plants exposed in the field in winter to SO_2. Nature, 299, 149-51.

Ball, D.J. (1976). Photochemical ozone in the atmosphere of Greater London. Nature, 263, 580-2.

Bell, J.N.B. (1982). Sulphur dioxide and growth of grasses. *In* Effects of Gaseous Air Pollution in Agriculture and Horticulture, eds. M.H. Unsworth & D.P. Ormrod, pp. 225-46. London: Butterworths.

Bell J.N.B., Rutter, A.J. & Relton, J. (1979). Studies on the effects of low levels of sulphur dioxide on the growth of *Lolium perenne* L. New Phytol., 83, 627-43.

Bennett, J.P. & Oshima, R.J. (1976). Carrot injury and yield response to ozone. J. Am. Soc. Hort. Sci., 101, 638-9.

Bennett, J.P., Oshima, R.J. & Lippert, L.F. (1979). Effects of ozone on injury and dry matter partitioning in pepper plants. Environ. Exp. Bot., 19, 33-9.

Bennett, J.P. & Runeckles, V.C. (1977). Effects of low levels of ozone on growth of crimsom clover and annual ryegrass. Crop Science, 17, 443-5.

Black, V.J., Ormrod, D.P. & Unsworth, M.H. (1982). Effects of low concentration of ozone, singly, and in combination with sulphur dioxide on net photosynthesis rates of *Vicia faba* L. J. Exp. Bot., 33, 1302-11.

Black, V.J. & Unsworth, M.H. (1979). Effects of low concentrations of sulphur dioxide on net photosynthesis and dark respiration of *Vicia faba*. J. Exp. Bot., 30, 473-83.

Blum, U. & Tingey, D.T. (1977). A study of the potential ways in which ozone could reduce root growth and nodulation in soybean. Atmos. Environ., 11, 737-9.

Bull, J.N. & Mansfield, T.A. (1974). Photosynthesis in

leaves exposed to SO and NO . Nature, <u>250</u>, 443-4.

Capron, T.M. & Mansfield, T.A. (1977). Inhibition of growth in tomato by air polluted with nitrogen oxides. J. Exp. Bot., <u>28</u>, 112-6.

Cox, R.A., Eggleton, A.E.J., Derwent, R.G., Lovelock, J.E. & Pack, D.H. (1975). Long-range transport of photochemical ozone in North-Western Europe. Nature, <u>255</u>, 118-21.

Davies, T. (1980). Grasses more sensitive to SO_2 pollution in conditions of low irradiance and short days. Nature, <u>284</u>, 483-5.

Davison, A.W. & Bailey, I.F. (1982). SO_2 pollution reduces the freezing resistance of ryegrass. Nature, <u>297</u>, 400-2.

Flagler, R.B. & Youngner, V.B. (1982). Ozone and sulphur dioxide effects on tall fescue: I. Growth and yield responses. J. Environ. Quality, <u>11</u>, 386-9.

Fowler, D. & Cape, J.N. (1982). Air pollutants in agriculture and horticulture. <u>In</u> Effects of Gaseous Air Pollution in Agriculture and Horticulture, eds. M.H. Unsworth & D.P. Ormrod, pp. 3-26. London: Butterworths.

Freer-Smith, P.H. (1984). The responses of six broadleaved trees during long-term exposure to SO_2 and NO_2. New Phytol., <u>97</u>, 49-61.

Hallgren, J.-E., Linder, S., Richter, A., Troeng, E. & Grant, L. (1982). Uptake of SO_2 in shoots of Scots pine: field measurements of net flux of sulphur in relation to stomatal conductance. Plant, Cell Environ., <u>5</u>, 75-83.

Harrison, R.M. & Holman, C.D. (1979). The contribution of middle- and long-range transport of tropospheric photochemical ozone to pollution at a rural site in North-West England. Atmos. Environ., <u>13</u>, 1535-45.

Jensen, K.F. (1981). Growth analysis of hybrid poplar cuttings fumigated with ozone and sulphur dioxide. Environ. Pollution (Series A), <u>26</u>, 243-50.

Jones, T. & Mansfield, T.A. (1982). The effects of SO_2 on growth and development of seedlings of <u>Phleum pratense</u> under different light and temperature environments. Environ. Pollution (Series A), <u>27</u>, 57-71.

Koziol, M.J. & Whatley, F.R. (1984). Gaseous air pollutants and plant metabolism. London: Butterworths.

Marshall, P.E. & Furnier, G.R. (1981). Growth responses of <u>Ailanthus altissima</u> seedlings to SO_2. Environ. Pollution (Series A), <u>25</u>, 149-53.

Mejstrik, V. (1980). The influence of low SO concentrations on growth reduction of <u>Nicotiana tabacum</u> L. cv. Samsum and <u>Cucumis sativus</u> L. cv. Unikat. Environ. Pollution (Series A), <u>21</u>, 73-6.

Oshima, R.J., Bennett, J.P. & Braegelmann, P.K. (1978).

Effect of ozone on growth and assimilate partitioning in parsley. J. Am. Soc. Hort. Sci., 103, 348-50.

Oshima, R.J., Braegelmann, P.K., Flagler, R.B. & Teso, R.R. (1979). The effects of ozone on the growth, yield, and partitioning of dry matter in cotton. J. Environ. Quality, 8, 474-9.

Parmeter, J.R. Jr. & Miller, P.R. (1968). Studies relating to the cause of decline and death of ponderosa pine in southern California. Plant Disease Reporter, 52, 707-11.

Reinert, R.A. & Weber, D.E. (1980). Ozone and sulphur dioxide-induced changes in soybean growth. Phytopathology, 70, 914-6.

Shimizu, H., Furukawa, A. & Totsuka, T. (1980). Effects of low concentrations of SO_2 on the growth of sunflower plants. Environ. Control Biol., 18, 39-47.

Siddiqui, M.Q. & Manners, J.G. (1971). Some effects of general yellow rust (*Puccinia striiformis*) infection on ^{14}carbon assimilation and growth in spring wheat. J. Exp. Bot., 22, 792-9.

Soderland, R. (1977). NO pollutants and ammonia emissions - a mass balance for the atmosphere over N.W. Europe. Ambio, 6, 118-22.

Tingey, D.T., Heck, W.W. & Reinert, R.A. (1971). Effect of low concentrations of ozone and sulphur dioxide on foliage, growth and yield of radish. J. Am. Soc. Hort. Sci., 96, 369-71.

Vukovich, F.M., Bach, W.D., Crissman, B.W. & King, W.J. (1977). On the relationship between high ozone in the rural surface layer and high pressure systems. Atmos. Environ., 11, 967-83.

Walmsley, L., Ashmore, M.R. & Bell, J.N.B. (1980). Adaptation of radish *Raphanus sativus* L. in response to continuous exposure to ozone. Environ. Pollution (Series A), 23, 165-77.

Wardlaw, I.F. (1968). The control and pattern of movement of carbohydrates in plants. Bot. Rev., 34, 79-105.

Whitmore, M.E. & Mansfield, T.A. (1983). Effects of long-term exposures to SO_2 and NO_2 on *Poa pratensis* and other grasses. Environ. Pollution (Series A), 31, 217-35.

Zeevaart, A.J. (1976). Some effects of fumigating plants for short periods with NO_2. Environ. Pollution, 11, 97-108.

THE WAY AHEAD: A CROP PHYSIOLOGIST'S VIEWPOINT

J.N. Gallagher
Lincoln College, Canterbury, New Zealand.

INTRODUCTION

The organisers of this symposium set three main aims for this paper: (i) to provide a view of the way in which various factors discussed during the symposium interact in controlling leaf growth in the field; (ii) to identify areas where further research is needed; and (iii) to assess the limitations of current investigatory approaches and suggest possible new approaches to an understanding of the control of leaf growth.

As a preliminary, I emphasise that the focus of this symposium is the control of leaf growth. An important distinction must be made between LIMITATIONS which arise from universal chemical and physical limits, and CONTROLS which have functional and adaptive values (Wareing, 1977). The Encyclopaedia Brittanica states that : "A control system is a means by which a variable quantity ... is made to conform to a prescribed norm ... systems may either hold the values of controlled quantities constant, or cause them to vary in a prescribed way". The growth of leaves is controlled so that leaf size, shape, orientation and physiological activities vary systematically with environment and ontogeny. The central theme of this paper is that in order to understand the control of leaf growth, we must elucidate the nature of the goals towards which the plant's control processes are directed.

Keeping in mind the prescribed aims, this paper is divided into four sections. In the first, I outline - albeit from a crop-physiological view point - the rationale for, and some of the objectives motivating research into leaf growth. This seems a necessary first step, one which enables current ideas and experimental work to be viewed in an appropriate perspective. In the second, I consider briefly, and with a philosophical slant, the various levels at which research into leaf growth may be done. In the third, I examine current experimental work in relation to the objectives motivating research and try to identify areas of ignorance where future endeavours may be fruitful. In the final section, I speculate on the type of research methodology that may be required to achieve a sound understanding of the control of leaf growth in the field.

RATIONALE AND OBJECTIVES

Rationale

The rationale for studies of leaf growth lies mainly in the importance of leaf area in determining plant growth - an importance which has long been appreciated (Gregory, 1921). Later work showed the primacy of leaf area over photosynthetic activity per unit leaf area in determining the productivity of plants and crops (e.g. Watson, 1952; Potter & Jones, 1977). More recent work has highlighted the direct relationship between the amount of radiation intercepted by the foliage of crops and crop growth rate and yield (Monteith & Elston, 1983). The slope of such relationships is often little affected by shortages of water or nutrients (see Chapter 12). Clearly, leaf growth is important to the establishment and maintenance of a foliage cover which will intercept most of the incoming solar radiation and produce fast growth and heavy yields. In addition the leaf area of plants is often a key determinant of their rate of transpiration. In many environments, a slowing of leaf growth with accompanying decreases on the number and size of leaves provides an effective means of restricting transpiration. The importance of leaf growth in determining both the growth and water use of plants is, therefore, hard to overestimate.

Objectives

From the foregoing it is clear that a prime objective of the crop physiologist and ecologist in studying leaf growth must be to understand the way in which internal and external factors interact in controlling the rate and duration of leaf expansion. This understanding is needed to form a basis for predicting the course of leaf expansion in different environments and the consequences that different patterns of leaf expansion have on growth and water use. To understand the factors controlling leaf growth, we must first understand the implications of different patterns of leaf growth on plant performance in a given environment, so that the "optimality" of different leaf growth strategies can be assessed. Leaf growth strategies might be aimed at achieving different and possibly conflicting goals, such as: the maximisation of total biomass; the maximisation of seed yield; the enhancement of life expectancy; the maximisation of vegetative spread and colonisation. Some insight on likely goals is a pre-requisite of a successful interpretation of observed patterns of leaf growth.

If the objective of research is to formulate principles underlying the control of leaf growth, then holistic aspects of plant response cannot be neglected. The physiological status and functioning of a whole plant depends not only on the size and activity of each of the individual plant organs, but also on their interaction (Fig. 1). Although environmental factors certainly have a major influence on leaf growth, intrinsic controls mean that leaf growth is intimately related with the growth of other plant organs.

LEVELS, UNDERSTANDING AND EXPLANATION

Experimentalists and theoreticians have recently drawn attention to the need to be aware of the levels of organisation that exist in biological organisms (Williams, 1975; Thornley, 1980). In this symposium studies of leaf growth have been reported at levels ranging from the community to the molecular. My main reason for introducing

Fig. 1. A simple scheme illustrating the interrelations between environment, the physiological activity and status of the plant and the growth rate and size of plant organs. Physiological "potential" refers to the maximum rates of photosynthesis and transpiration per unit area for leaves, and the maximum rates of nutrient and water uptake per unit of surface area for roots.

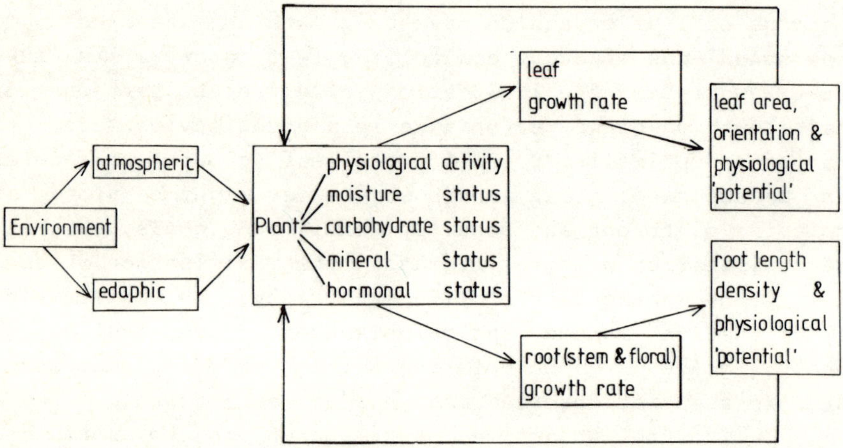

the topic of "levels" has been succinctly stated by Thornley (1980): "A good description of a phenomenon at a given level in the hierarchy not only gives a firm stepping off point for a reductionist approach, but it allows the person working at the next higher level to synthesise an understanding of his problem". It is important, therefore, to be aware of the objectives of research in determining the level at which to proceed.

I shall make two points on the importance of levels to the study of leaf growth. My points will be demonstrated in the context of a simple identity expressing the leaf area of a plant (A_p) as the product of several components representing lower levels of organisation. Thus:

$$A_p = N_b \times N_l \times \bar{A} \quad (1)$$

where N_b is the number of branches per plant, N_l is the number of leaves per branch, and $\bar{A}$ is the mean area per leaf. The variable $\bar{A}$ could be further described as a function of the average number of cells per leaf, their arrangement and size and so on.

My first point is prompted by the observation that the majority of papers in this symposium have concentrated on or below the level of the leaf. Yet from equation (1) it is evident that for an understanding at the level of the organism, we need a description of the way in which various factors influence the demographic processes governing branching and leaf numbers as well as leaf size. In fact, for *Linum usitatissimum* Bazzaz & Harper (1977) found leaf number per plant to be the major determinant of leaf area: for some grasses and cereals, the number of tillers (branches) per plant is the major determinant of leaf area. To assess the relevance and importance of research into a particular process at a particular level it is therefore necessary to explore quantitatively the inter-relations of the selected process with processes at the same and at adjacent levels of the hierarchy.

My second point follows from Thornley's (1980) viewpoint stated earlier. It is that if we can describe the effect of physical and physiological factors on the terms on the right of equation (1), then it should be possible to predict the response of leaf growth to these factors and to understand the nature of the response at the level of the organism. It may, of course, be impossible to specify with sufficient exactness some of the transfer functions for the black boxes inherent in this approach. Then, investigations at a lower level in the hierarchy in search of a better description or explanation will be needed. However, the objectives of the lower level investigations must be remembered. Two questions might usefully be kept in mind. Has the work so far improved understanding? Does the problem look soluble and if not, can another approach be taken? The principle underlying these questions might be termed "restrained reductionism": any tendency to delve lower in the hierarchy is restrained by keeping the objectives of the original level of interest firmly in mind.

EXPERIMENTATION AND ANALYSIS: DIRECTION AND NEEDS

Thompson (1981) distinguished four phases in the scientific study of a natural phenomenon. The first phase consists largely of gathering information and facts about the phenomenon. Experimental work during this phase is typically aimed at getting a general idea as to which factors have important effects on the phenomenon and which interactions between factors are significant. Analysis during this phase is typically done using general statistical response models. *A posteriori* regression procedures are also often used in an attempt to identify variables having an important influence on the phenomenon. The work presented at this symposium shows that studies on leaf growth have largely shifted to a second phase, characterised by the quantitative formulation of principles and empirical laws. Analysis in this phase involves the application of specific response models, many of which have a mechanistic basis. The number of factors studied may be limited, but each factor is often applied at many levels. I believe that the shift away from 2^n or 3^n factorial experiments is to be welcomed, indicating as it does, a greatly improved knowledge of the systems under study.

In the field of leaf growth, a salutary reminder of the difference between the two phases of study outlined above appeared recently (Haun & Coston, 1983). The authors were concerned to determine the factors controlling the rate of leaf unfolding of peach trees. To do this: "A total of 108 independent variables were included in stepwise or multiple regressive analyses performed with computer programs ...". The upshot of the ensuring orgy of electronic calculations was a regression equation with 10 predictor variables, the physical or physiological relevance of many of which was hard to see. The equation explained about 65%

of the variation in leaf appearance rate in the season for which it was fitted. However, in an independent test of another season's results, for most of the time the assumption of a constant rate of leaf unfolding gave a better fit to the observations than did predictions from the regression equation. This example emphasises the need for a better understanding of the influences of environmental and physiological factors on the process leading in turn to a mechanistic model capable of more accurate and more general predictions.

The construction of a model represents a third phase of study: one in which mathematical models are formulated and validated. Studies on leaf growth appear to be on the threshold of this third phase. The potential importance and implications of this third phase for our understanding of the control of leaf growth are considered in the Modelling and Synthesis section of this chapter.

The remainder of this section concerns specific areas where I see a need for more or better information about leaf growth. I first consider the need to be aware of, and to monitor, relevant features of the plant's physical environment. I then outline the need to investigate the factors controlling the initiation and early growth of leaves. Next, I try to identify gaps in our knowledge about the influence of certain environmental factors during the main period of leaf expansion. Finally, I stress the need to study both the inter-relations between the growth of leaves and other plant organs, and the influence of growth conditions on the subsequent physiological activity of leaves.

Microclimate

Almost 30 years ago, at another symposium on leaf growth, Penman (1956) pointedly drew attention to the failure of physiologists to make relevant microclimatic measurements. Since then, most physiologists have become aware of the need to measure appropriate microclimatic variables (Marshall, 1984). None the less, there is still room for improvement. A particular problem is the measurement of the spectral composition of radiation, a problem which Penman (1956) perspicaciously foresaw. It must also be recognised that in the field, microclimatic variables such as soil and air temperatures, radiation and humidity are coupled to varying degress. The temperature of the soil where roots are extracting water may differ by more than 10 C from leaf temperature. Daylength and insolation move in phase, followed with a lag by temperature. Although the abrupt changes in radiation experienced in growth chambers may have little influence on photosynthesis (Wilson et al., 1978), account must be taken of the absence of end-of-day effects (see for example Milford & Lenton, 1976) and the frequent absence of correlations between temperature and insolation. This is especially true when the objective of research is to determine the influence on growth of a single microclimatic factor isolated from the complex present in the field.

Early Ontogeny and Predetermination

The rate and duration of leaf primordium initiation determine the number of leaves a shoot will produce. Recent quantitative descriptions of the initiation process mark an important advance. Kirby (1974) showed that a simple two-segment linear model adequately described the temporal trend of primordium initiation in spring wheat. Kirby's model has since been shown to be applicable to barley (Kirby & Riggs, 1978) and to winter wheat when time is corrected for temperature (Gallagher, 1978). By enabling final leaf number to be analysed as the product of a rate and duration of primordium initiation, the model is already proving useful in understanding environmental and genotypic control of leaf number (Kirby & Riggs, 1978; Baker & Gallagher, 1983; Warrington & Kanemasu, 1983).

There seems little doubt that events occuring while a leaf is still small can be important in determining final leaf size (Milthorpe, 1956; Milford & Riley, 1980). The idea is not new. Balls (1914) stated: "... we are largely in error when we seek to ascertain the causes of fluctuation in some feature of an organism by integrating the effects of those environmental factors which were operating at, or nearly before, the time when the feature in question was manifest ... such factors merely exert a subsidiary deforming influence on a scaffolding which was predetermined ... often at a much earlier stage ...". For plants in which particular leaves provide much of the assimilate for fruit or seed growth it is important to know the extent to which the size of these leaves is predetermined - the rigidity of the early scaffolding. Working with rice, Yamazaki (1963, 1964) showed that the vascularisation and width of a leaf were determined by the width of the apex at initiation and by conditions during the early stages of growth. I believe that in the study of predetermination more information is needed about the relative importance of cell number and cell size in determining the size of the mature leaf.

The intricate and highly related patterns of growth exhibited by primordial leaves before they unfold (or appear) undoubtedly has some bearing on the possible predetermination of leaf size. Recently, patterns of growth have been analysed quantitatively and deductions made about their relations with apical dome size and the size of mature leaves (Horie et al., 1979; Pieters, 1983). However, these studies have generally been made in stable environmental conditions. The effects of perturbations on these growth patterns caused by environmental and hormonal factors are largely unknown. Vendeland et al., (1982) showed that drought could severely disrupt the regular pattern of leaf expansion on the apex of soya bean, causing the plastochron index to vary. The implications of this disruption on final leaf size were not, however, reported.

The determination of mathematical functions describing temporal trends of leaf size is an area in which the application of computers and statistical packages looks appropriate and useful. Subsequent analysis can compare the influence of treatments on well-defined growth variables

(Causton et al., 1978). A comprehensive application of this functional approach enables a thorough description of the influence of environmental and ontogenetic factors on leaf growth to be established (Jones & Hesketh, 1980).

Seasonal, Photoperiodic and Photomorphogenetic Effects

Season and photoperiod affect ontogeny. They often strongly influence the number of leaves that a shoot will produce, the appearance rate of leaves and final leaf size (Koller & Kiegel, 1972; Kirby et al., 1982). For example, it appears that in cereal species the thermal rate of leaf appearance is set soon after emergence, possibly by the rate of change of daylength at the time of emergence (Baker et al., 1981; Hay & Tunncliffe-Wilson, 1982). It seems important that the effects of such early influences on the rates of leaf initiation and appearance are clearly understood if subsequent patterns of leaf expansion are to be correctly interpreted.

Photoperiod, together with many other photomorphogenetic stimuli, is probably detected by phytochrome (Vince-Prue & Tucker, 1983). Photomorphogenetic stimuli can strongly influence growth of plants and leaves in the field (Milford & Lenton, 1976; Morgan & Smith, 1979). It is therefore disconcerting to read in a recent review of photomorphogenetic influences on leaf growth: "many investigations of the photocontrol of leaf development have been concerned solely with plants grown in the dark" (Vince-Prue & Tucker, 1983). This investigational style must surely change during the next decade. Obviously, it is important for plant growth and survival that leaf growth responds in an appropriate manner to conditions of shade and to competition for light. There is, I believe, a great need for a better understanding and explanation of the responses of leaf growth to photomorphogenetic factors in the field. To meet this need, we must be able to measure the appropriate spectral ratios. Hopefully, recent advances in instrumentation will continue so that these measurements will be possible (e.g. Woodward, 1983).

Temperature

Temperature can exert a strong influence on the rate and duration of leaf initiation, the rate of leaf appearance and the size of mature leaves (Dale & Milthorpe, 1983). However, as yet, there seems to have been little attempt to comprehensively study the effect of temperature on these processes, let alone to determine the effects of temperature and photoperiod acting in concert in the field. Chao & Loomis (1947) showed that leaf expansion responded strongly to temperature from which they inferred that chemical rather than physical processes limited leaf expansion. They also found that the Q_{10} for expansion processes declined with temperature, - some of the responses they recorded appear linear over a considerable range of temperature. We are still unsure about the nature of the response of leaf expansion rate to temperature. The concept of a thermal rate, advocated by workers from Sutton Bonington, depends on there being a linear response of expansion rate to temperature. Is then, the Arrhenius plot a "conspiracy" as suggested in a previous SEB meeting, or is it a valid means of describing and analysing the influence of temperature on the physiological processes involved with leaf growth (e.g. Kemp, 1980)? Is the stochastic model advanced by Sharpe & De Michelle (1977) likely to be useful, and are its parameters biochemically meaningful? At what base temperatures do leaf growth and developmental processes stop? Are these base temperatures fixed or do they vary as Kirby et al. (1982) suggest is the case for the leaf appearance rate of barley, and as is the case for other physiological processes (Larcher, 1980)? Some results suggest that the effects of temperature on cereal leaf extension rate and duration are compensatory - resulting in stable leaf size (Gallagher et al., 1978; Dale, 1983). Other results showed a marked dependence of cereal leaf size on temperature (Friend, 1966). All of these questions and contradictions are important and must be investigated before we can understand and explain the influence of temperature on leaf size.

Drought, Water Stress and Mineral Nutrition

It is evident that we are now able to explain more clearly how water stress can influence leaf growth. At one level the nature of the interactions between temperature and the saturation deficit of the atmosphere are becoming clearer (Squire *et al*., 1983). (It is interesting to note that Loomis in 1934 made similar measurements to those of Squire *et al*. and made a similar interpretation, albeit using cruder procedures.) The study of the mechanisms underlying the response of leaf growth to water stress has been helped by the development of equipment which enables the components of tissue water potential and cell wall extensibility to be measured (*e.g*. Van Volkenburgh *et al*., this volume). I hope that future work will explain the different responses of leaf growth to turgor that have been observed (Wenkert *et al*., 1978; Gallagher *et al*., 1978; Squire *et al*., 1983).

The relevance of various measures of plant water status made in various tissues to the physiological processes involved in leaf growth was critically discussed during the symposium. Contributions by Hsiao, Steudle and Tomos give a lucid account of the problems of making accurate and relevant measurements of tissue water status. Failure to address these problems probably accounts for some of the differences observed in the response of leaf extension to turgor discussed above. It is also important to remember that despite the fascinating nature of an observed response at the cellular or sub-cellular level, that response may be of little significance to the functioning of the whole plant. For instance, in a recent study Wilson & Ludlow (1983) found considerable osmotic adjustment in the leaves of three tropical grasses in response to drought. However, they concluded that in their conditions the contribution of this osmotic adjustment to the maintenance of leaf growth during drought was "small". This clearly demonstrates the need to quantitatively assess the implications of an intriguing physiological response operating at a low level, to the functioning of an organism at a higher level.

The possibility that tissue water stress *per se* may not be the primary factor limiting leaf growth during a drought in the field is intriguing. A dry soil usually means a poor availability of mineral nutrients for growth. Garwood & Williams (1967) demonstrated this effect experimentally for grass. Since then, the frequently overriding importance of limiting supplies of soil nutrients to pasture growth, even in arid regions, has been amply demonstrated (*e.g.* Penning de Vries, 1982). The interactions between water and nutrient supplies need elucidating further, especially as it appears that nutrients can influence the hydraulic conductivity of leaf tissues (Radin & Boyer, 1982).

Drought-nutrient interactions aside, there is an astonishing paucity of information about the influence of mineral nutrients on leaf growth. This is despite the fact that it is well known that supplies of major nutrients often limit leaf size. For instance, shortage of nitrogen can slow the rates of leaf initiation and appearance and decrease the rate of cell division in leaf meristems (Dale & Wilson, 1978; Volenec & Nelson, 1983; Steer & Hocking, 1983). Nitrogen can also have massive effects on the duration and rate of leaf area expansion and final leaf area (Steer & Hocking, 1983; Thomas, 1983). In addition, nitrogen and other nutrients can profoundly influence leaf structure and photosynthetic activity (*e.g.* Osman & Milthorpe, 1971; Field & Mooney, 1983). The current level of ignorance about the influence of mineral nutrition on leaf growth represents a grave imbalance in urgent need of redress.

Insolation and Assimilate Supply

Bursts of strong sunshine have long been known to slow leaf extension and this has been associated with tissue water deficits (Thut & Loomis, 1944). Recent work has confirmed these responses (Parish & Wolfe, 1983). In addition to these short term effects on leaf growth modulated through changes in tissue water status, irradiance may also influence leaf growth through its effect on assimilate supplies (Kemp, 1981; Takami et al., 1981). In grasses and temperate cereals it seems unlikely that soluble sugar levels limit leaf extension rate in the field as there is evidence of considerable buffering capacity (Kemp, 1981; Sambo, 1983). However, it should be noted that some experimental work suggests that reserve carbohydrates are not always available for bud and leaf growth in grasses (McIntyre, 1977). There may therefore be an important need to determine the concentration of specific sugars in this kind of work. It seems that the influence of soluble carbohydrates of the growth of dicotyledenous leaves awaits study.

The influence of insolation during growth on the physiological activities of leaves can be considerable (Charles-Edwards & Ludwig, 1974). Most evidence supports the hypothesis that the amount of photosynthetic machinery developed in expanding leaves depends on the irradiance of the plant during growth (see Chapter 10). However, Bunce (1983) convincingly demonstrated that there is a need to take account of not only the irradiance of a leaf during growth, but also its expansion rate. He found that maximum photosynthetic rate increased as the ratio of daily net photosynthesis to leaf expansion became larger.

Interrelations between Leaf Growth and Plant Growth

If we are to understand the control of leaf growth it seems essential to understand the ways in which the growth and activity of leaves are related to the growth and activity of other plant organs (Fig. 1). The interplay between leaf and root growth is particularly important for two reasons. First, the roots provide the water needed to maintain turgor in the leaf and the mineral nutrients needed for growth. There is sound evidence that a "functional equilibrium" exists between root and shoot growth (Brouwer, 1966). Thus if water is in short supply, the ratio of root growth to shoot growth increases. This type of response has recently been convincingly demonstrated in the field (Huck et al., 1983) and poses interesting questions about the co-ordination of root and shoot growth. Does the functional equilibrium depend primarily on a steepening gradient of soluble sugars from the shoot to the root, or does it depend on hormonal control? It is clear that the balance between supplies of photosynthate from the leaves and water and minerals from the roots may be an important variable influencing the control and pattern of leaf growth. The remarkable effect of gaseous pollutants in decreasing root:shoot ratios awaits exploration (see Chapter 14).

The second reason why leaf-root interrelations are important involves the capacity of roots for hormone synthesis. If root growth is restricted in some way, shoot and leaf growth are also likely to be decreased. Richards (1981) suggested that the co-ordinated response was mediated via changes in the synthesis of hormones (particularly cytokinins) in the roots. Recent work in the Netherlands has also implicated active root growth as an important determinant of the leaf growth of grasses in the spring (B. Deinum, pers. comm.). The consequences of ignoring the root when studying the shoot should not be overlooked.

The role of hormones in interrelations between root and shoot and the general co-ordination of growth is obviously worth investigating - especially as this may provide a means of manipulating plant growth to man's advantage. Current endeavours in this field suggest that the way ahead will not be easy (Chapter 3). Nonetheless, because the possibility of controlling plant growth chemically is an alluring one, research in this area will proceed - probably via the pharmacological approach (Goodwin & Erwee, 1983).

To understand the physiological basis for control of leaf growth the effects of internal and external factors on the physiological activity of leaves can evidently be just as important as their influences on size alone. For instance, a simple measure of size,such as area or length, is an inadequate expression of the resultant of the various leaf growth processes when it is known that photosynthetic activity may vary by a factor of four (Bunce, 1983) and water use efficiency by a factor of two (Nobel, 1980). The proliferation of work based solely on measurements of leaf growth made using auxanometers is, therefore, a possible cause for concern. There is a need to study the growth in size of leaves in relation to their physiological activity, the number and microclimate of other leaves on the plant, and the extent and activity of the root system. The last section examines some ways in which this integration may be achieved.

MODELLING AND SYNTHESIS: DIRECTIONS OF THE FUTURE?

Philosophy

The widespread application of mathematical modelling to the plant sciences was originally greeted with much scepticism. Nowadays, the modelling approach is accepted by many as matter-of-fact and essential to progress in both understanding and explaining plant behaviour. In essence, I take the view that: "Our simplified picture of reality is often called a 'model'; it leaves out a lot of the complex details of reality, but we hope that it can describe the behaviour closely enough to be useful, and to give us a feeling of understanding the real phenomenon better" (Goldstein & Goldstein, 1978). For the physiologist, modelling is the only way of synthesising the details - the bits and pieces of knowledge that result from reductionist labours - into a representation of the behaviour of the whole, be it a cell or an organism or a community. This synthesis is surely the aim of plant and crop physiologists.

To construct models which will aid in our understanding of the control of leaf growth and the implications of this control to whole plants an communities of plants, we do not need information merely about leaf size. As stressed earlier, we also need information on the physiological activities of leaves and their chemical composition, on the interrelations between the growth of leaves and other organs, and on the plant environment. I believe that we need to take this knowledge and place it in the quantitative conceptual framework of a mathematical model. Such a model may also contain conjectures about mechanisms underlying responses and interrelationships among processes. The model can then be used to explore the consequences of different control rules on the growth, reproductive ability and life expectancy of plants. In so doing, the study of leaf growth has moved into the fourth phase of investigation described by Thompson (1981): "use of mathematical models to gain scientific insight". Thus a model could be used to help determine optimal strategies for leaf growth in different environments. Hypotheses resulting from such studies must then be tested experimentally.

In discussing the use of models I have introduced an element of teleology which some find deeply disturbing (e.g. Kramer, 1980). Horn (1979) has suggested that there is a respectable place for teleological thought in biology "in the framing of questions rather than in the answering of them". It is from the point of view of hypothesis formulation that teleology is introduced here. I shall illustrate the model-building approach with some examples given in increasing order of complexity.

Examples

The physical factors governing the energy balance of leaves, and the physiological factors governing the photosynthesis, respiration and transpiration of leaves are well documented (e.g. Larcher, 1980). Using this knowledge it is possible to deduce the sizes, shapes and orientation of foliage organs suited both to maintaining tissue temperatures within physiologically acceptable limits, and to achieving an appropriate balance between photosynthesis, respiration and transpiration in different environments.

The approach described above for leaves can be extended so as to deduce how the whole plant may respond to diverse environmental conditions. Several groups of workers in the U.S.A. have been active in this area (e.g. Orians & Solbrig, 1977; Mooney & Gulman, 1979). The arguments and conjectures of these groups are expressed within the framework of a type of physiological cost-benefit analysis. Essentially three sets of costs are distinguished: (i) a construction cost, the quantity of carbon and minerals required to grow the leaf; (ii) a support and supply cost, the carbon and minerals required to support the leaf with petiole and branch structures, and to transport water and nutrients to and from the leaf *via* an adequate vascular system and; (iii) a maintenance cost, the energy (in terms of carbon) needed to ensure the structural integrity and continued functioning of plant tissues. These "costs" have to be balanced by "gains" from leaves measured in terms of carbon fixation. The nub of the hypothesis is that the marginal cost of expenditure of a resource on leaf tissue equals the marginal gain from growing that tissue. The resulting optimal growth pattern is one that maximises net carbon gain. Maximising net carbon gain is admittedly a simple goal to suggest for the control of plant growth. However, the U.S.A. groups consider that the degree of success in attaining this goal is a key measure of plant performance and competitive ability.

A simple example of the implications of the marginal cost versus marginal gain hypothesis is to be found by considering the growth of plants in weak light, for example in the understory of a forest. In such a microclimate the production of thick leaves, rich in photosynthetic enzymes, capable of high rates of photosynthesis in bright sunshine and associated with high rates of respiration would lead to net carbon loss rather than maximal carbon gain. To maximise net carbon gain a more appropriate growth response would be to grow thin leaves with low maximum rates of photosynthesis achieved in weak light and slow respiration. As Lichtenthaler (Chapter 10) shows, the leaf growth of species in weak light is indeed controlled to achieve these physiological and morphological characteristics.

Gulman & Chu (1981) made a direct experimental test of the "maximal carbon gain" hypothesis in relation to the insolation and nitrogen supply of Chaparal shrub (Diplaucus aurantiacus). Their results supported the idea that leaf growth and partitioning were controlled to maximise net carbon gain. A similar concept of optimality has been advanced by Schulze et al. (1983) for cowpea (Vigna unguiculata). They postulated that "carbohydrate partitioning should lead to maximum accumulation of biomass without a decrease in leaf water status under a range of conditions of evaporative demand". Their model involved a special development of optimal control theory and a simplified treatment of leaf and root growth and root water uptake as functions of soil water content. The predictions of the model were compared with the results of an experiment in which plants were grown at two levels of atmospheric humidity and soil water content. The experiment bore out the validity of the optimality principles of the model.

CONCLUSION

Clearly, many of these model-based attempts to elucidate the principles which determine the control of leaf and plant growth in different environments are in their infancy. Yet I believe that they mark an essential first step with important implications for the present meeting. Only when we combine knowledge about leaf growth and activity with knowledge about the rest of the plant can we begin to understand the implications of different possible responses of leaf growth to extrinsic and intrinsic factors in the field. Only by considering the whole plant can we identify the performance criteria which the control mechanisms associated with leaf growth may be trying to satisfy.

ACKNOWLEDGEMENT

I thank J.B. Reid and C.A. Macken for helpful comments on an early draft of this paper.

REFERENCES

Baker, C.K. & Gallagher, J.N. (1983). The development of winter wheat in the field. 2. The control of primordium initiation rate by temperature and photoperiod. J. Agric. Sci., Camb., 101, 337-344.

Baker, C.K., Gallagher, J.N. & Monteith, J.L. (1981). Daylength change and leaf appearance in wheat. Plant, Cell and Environ., 3, 285-287.

Balls, L. (1914). Pre-determination of fluctuation. Proc. Camb. Phil. Soc., Math. Phy. Sci., 17, 469-470.

Bazzaz, F.A. & Harper, J.L. (1977). Demographic analysis of the growth of Linum usitatissimum. New Phytol., 78, 193-208.

Brouwer, R. (1966). Root growth of grasses and cereals. In The Growth of Cereals and Grasses, ed. F.L. Milthorpe and J.D. Ivins, pp. 153-166. London: Butterworths.

Bunce, J.A. (1983). Photosynthetic characteristics of leaves developed at different irradiances and temperatures: an extension of the current hypothesis. Photosyn. Res., 4, 87-97.

Causton, D.R., Elias, C.O. & Hadley, P. (1978). Biometrical studies of plant growth. I. The Richard's function and its application in analysing the effects of temperature on leaf growth and Environ., 1, 163-184.

Chao, M.D. & Loomis, W.E. (1947). Temperature coefficients of cell enlargement. Bot. Gaz., 225-231.

Charles-Edwards, D.A. & Ludwig, L.J. (1974). The basis of expression of leaf photosynthetic activities. In Environmental and Biological Control of Photosynthesis, ed. R. Marcelle, pp. 37-44. The Hague: Junk.

Dale, J.E. (1983). Some effects of temperature and irradiance on growth on the first four leaves of wheat, Triticum aestivum. Ann. Bot., 50, 851-858.

Dale, J.E. & Milthorpe, F.L. (1983). General features of the production and growth of leaves. In The Growth and Functioning of leaves, eds. J.E. Dale and F.L. Milthorpe, pp. 151-178. Cambridge: Cambridge University Press.

Dale, J.E. & Wilson, R.G. (1978). A comparison of leaf and ear development in barley cultivars as affected by nitrogen supply. J. Agric. Sci. Camb., 90, 503-508.

Field, C. & Mooney, H.A. (1983). Leaf age and seasonal effects on light, water and nitrogen use efficiency in a California shrub. Oecologia, 56, 348-355.

Friend, D.J. (1966). The effects of light and temperature on the growth of cereals. In The Growth of Cereals and Grasses, eds. F.L. Milthorpe and J.D. Ivins, pp. 181-199. London: Butterworths.

Gallagher, J.N. (1978). Field studies of cereal leaf growth. I. Initiation and expansion in relation to temperature and ontogeny. J. Exp. Bot., 30, 625-636.

Gallagher, J.N., Biscoe, P.V. & wallace, J.S. (1978). Field studies of cereal leaf growth. IV. Winter wheat leaf extension in relation to temperature and leaf water status. J. Exp. Bot., 30, 657-668.

Garwood, E.A. & Williams, T.E. (1967). Growth, water use and nutrient uptake from the subsoil by grass swards. J. Agric. Sci. Camb., 69, 125-130.

Goldstein, I.F. & Goldstein, M. (1978). How We Know: An Explanation of the Scientific Process. New York: Plenum Press.

Goodwin, P.F. & Erwee, M.G. (1983). Hormonal influences of leaf growth. In The Growth and Functioning of Leaves, eds. J.E. Dale and F.L. Milthorpe, pp. 207-232. Cambridge: Cambridge University Press.

Gregory, F.G. (1921). Studies in the energy relations of plants. I. The increase in area of leaves and leaf surface of Cucumis sativus. Ann. Bot., 35, 93-123.

Gulman, S.L. & Chu, C.C. (1981). The effects of light and nitrogen on photosynthesis, leaf characteristics, and dry matter allocation in the Chaparral shrub, Diplaucus aurantiacus. Oecologia, 49, 207-212.

Haun, J.R. & Coston, D.C. (1983). Relationship of daily growth and development of peach leaves and fruit to environmental factors. J. Amer. Soc. Hort. Sci., 108, 666-671.

Hay, R.K.M. & Tuncliffe-Wilson, G. (1982). Leaf appearance and extension in field-grown winter wheat plants: the importance of soil temperature during vegetative growth. J. Agric. Sci., Camb., 99, 403-410.

Horie, T., De Wit, C.T., Goudriaan, J. & Bensink, J. (1979). A formal template for the development of cucumber in its vegetative state. Proc. Roy. Acad. Sci. Neth., 82, 433-479.

Horn, H.S. (1979). Adaptation from the Perspective of Optimality. In Topics in Plant Population Biology, eds. O.T. Solbrig, S. Jain, G.B. Johnson and P.H. Raven, pp. 48-61. New York: Columbia University Press.

Huck, M., Ishihara, K., Peterson, C.M. & Ushijima, T. (1983). Soybean adaptation to water stress at selected stages of growth. Plant Physiol., 73, 422-427.

Jones, J.W. & Hesketh, J.D. (1980). Predicting leaf expansion. In Predicting Photosynthesis for Ecosystem Models: Volume II, eds. J.D. Hesketh and J.W. Jones, pp. 85-122. Boca Raton: CRC Press.

Kemp, D.R. (1980). The growth rate of successive leaves of

wheat plants in relation to sugar and protein concentrations in the extension zone. J. Exp. Bot., 31, 1399-1411.
Kemp, D.R. (1981). The growth rate of wheat leaves in relation to the extension zone sugar concentration manipulated by shading. J. Exp. Bot., 32, 141-150.
Kirby, E.J.M. (1974). Ear development in spring wheat. J. Agric. Sci. Camb., 82, 437-447.
Kirby, E.J.M. & Riggs, T.J. (1978). Developmental consequences of two-row and six-row ear type in spring barley. 2. Shoot apex, leaf and tiller development. J. Agric. Sci., Camb., 91, 207-216.
Kirby, E.J.M., Appleyard, M. & Fellowes, G. (1982). Effect of sowing date on the temperature response of leaf emergence and leaf size in barley. Plant, Cell and Environ., 5, 477-484.
Koller, D. & Kiegel, J. (1972). The growth of leaves and tillers in *Oryzopsis miliacea*. In The Biology and Utilisation of Grasses, eds. V.B. Youngner and C.M. McKell, pp. 115-134. New York: Academic Press.
Kramer, P.J. (1980). Drought, stress and the origin of adaptations. In Adaptation of Plants to Water and High Temperature Stress, eds. N.C. Turner and P.J. Kramer, pp. 7-20. New York: John Wiley.
Larcher, W. (1980). Physiological Plant Ecology (2nd Edition). Berlin: Springer-Verlag.
Loomis, W.E. (1934). Daily growth of maize. Amer. J. Bot., 21, 1-6.
McIntyre, G.I. (1977). The role of nutrition in apical dominance. In Integration of Activity in the Higher Plant, ed. D.H. Jennings, pp. 251-273. Cambridge: Cambridge University Press.
Marshall, B. (1984). Proceedings of SEB Symposium on Environmental Measurement (In Press).
Milford, G.F.J. & Lenton, J.R. (1976). Effect of photoperiod on growth of sugar beet. Ann. Bot., 40, 1309-1315.
Milford, G.F.J. & Riley, J. (1980). The effects of temperature on leaf growth of sugar beet varieties. Ann. Appl. Biol., 94, 431-443.
Milthorpe, F.L. (1956). The relative importance of the different stages of leaf growth in determining the resultant area. In The Growth of Leaves, ed. F.L. Milthorpe, pp. 141-148. London: Butterworths.
Monteith, J.L. & Elston, J. (1983). Performance and productivity of foliage in the field. In The Growth and Function of Leaves, eds. J.E. Dale and F.L. Milthorpe, pp. 499-518. Cambridge: Cambridge University Press.
Mooney, H.A. & Gulman, S.L. (1979). Environmental and evolutionary constraints on the photosynthetic characteristics of higher plants. In Topics in Plant Population Biology, eds. O.T. Solbrig, S.

Jain, G.B. Johnson and P.H. Raven. New York: Columbia University Press.

Morgan, D.C. & Smith, H. (1979). A systematic relationship between phytochrome-controlled development and species habitat, for plants grown in simulated natural radiation. Planta, 145, 253-258.

Nobel, P.S. (1980). Leaf anatomy and water use efficiency. In Adaptation of Plants to Water and High Temperature Stress, eds. N.C. Turner and P.J. Kramer, pp. 43-55. New York: Wiley Interscience.

Orians, G.H. & Solbrig, O.T. (1977). A cost-income model of leaves and roots with special reference to arid and semi-arid areas. Amer. Nat., 111, 677-690.

Osman, A.M. & Milthorpe, F.L. (1971). Photosynthesis of wheat leaves in relation to age, illuminance and nutrient supply. II. Results. Photosynthetica, 5, 61-70.

Parish, D.J. & Wolfe, D.D. (1983). Kinetics of tall fescue leaf elongation: response to changes in illumination and vapour pressure. Crop Sci., 23, 659-663.

Penman, H.L. (1956). Discussion. In The Growth of Leaves, ed. F.L. Milthorpe, pp. 168-169. London: Butterworths.

Penning de Vries, F.W.T. (1982). Systems analysis and models of crop growth. In Simulation of Plant Growth and Crop Production, eds. F.W.T. Penning de Vries and H.H. Van Laar, pp. 9-19. Wageningen: Pudoc.

Pieters, G.A. (1983). Growth of Populus euramericana. Physiol Plant., 57, 455-462.

Potter, J.R. & Jones, J.W. (1977). Leaf area partitioning as an important factor in growth. Plant Physiol., 59, 10-14.

Radin, J.W. & Boyer, J.S. (1982). Control of leaf expansion by nitrogen nutrition in sunflower plants. Plant Physiol., 69, 771-775.

Richards, D. (1981). Root-shoot interactions in fruiting tomato plants. In Structure and Function of Plant Roots, eds. R. Brouwer, O. Gasparikova, J. Kolek and B.C. Loughman. pp. 373-380. The Hague: Martinus Nijhof.

Sambo, E.Y. (1983). Leaf extension rate in temperate pasture grasses in relation to assimilate pool in the extension zone. J. Exp. Bot., 34, 1281-1290.

Schulze, E.D., Schilling, K. & Nagarajah, S. (1983). Carbohydrate partitioning in relation to whole plant production and water use of Vigna unguiculata (L.) Walp. Oecologia, 58, 169-177.

Sharpe, P.J.H. & De Michelle, D.W. (1977). Reaction kinetics of poikilotherm development. J. Theor. Biol., 64, 649-670.

Squire, G.R., Black, C.R. & Ong, C.K. (1983). Response to

saturation deficit of leaf extension in a stand of pearl millet (Pennisetum typhoides S. & H.). II. Dependence on leaf water status and irradiance. J. Exp. Bot., 34, 856-865.
Steer, B.T. & Hocking, P.J. (1983). Leaf and floret production in sunflower (Helianthus annus L.) as affected by nitrogen supply. Ann. Bot., 52, 267-277.
Takami, S., Rawson, H.M. & Turner, N.C. (1981). Leaf expansion of four sunflower (Helianthus annus L.) cultivars in relation to water deficits. II. Diurnal patterns during stress and recovery. Plant, Cell and Environ., 5, 279-286.
Thomas, H. (1983). Analysis of the nitrogen response of leaf extension in Lolium temulentum seedlings. Ann. Bot., 51, 363-371.
Thompson, M. (1981). Mathematisation in the Sciences. In Mathematics Tomorrow, ed. L.A. Steen, pp. 243-250. New York: Springer-Verlag.
Thornley, J.H.M. (1980). Research strategy in the plant sciences. Plant, Cell and Environ., 3, 233-236.
Thut, H.F. & Loomis, W.E. (1944). Relation of light to growth of plants. Plant Physiol., 19, 117-130.
Vendeland, J.S., Sinclair, T.R., Spaeth, S.C. & Cortes, P.M. (1982). Assumptions of plastochron index: evaluation with soyabean under field drought conditions. Ann. Bot., 50, 673-680.
Vince-Prue, D.O. & Tucker, D.J. (1983). Photomorphogenesis in leaves. In The Growth and Functioning of Leaves, eds. J.E. Dale and F.L. Milthorpe, pp. 233-269. Cambridge: Cambridge University Press.
Volenec, J.J. & Nelson, C.J. (1983). Responses of tall fescue leaf meristems to N fertilisation and harvest frequency. Crop Sci., 23, 720-724.
Wareing, P.F. (1977). Growth substances and integration in the whole plant. In: Integration of Activity in the Higher Plant, ed. D.H. Jennings, pp. 337-365. Cambridge: Cambridge University Press.
Warrington, E.J. & Kanemasu, E.T. (1983). Corn growth response to temperature and photoperiod. II. Leaf initiation and leaf appearance rates. Agron. J., 74, 755-761.
Watson, D.J. (1952). The physiological basis of variation in yield. Adv. Agron., 4, 101-145.
Wenkert, W., Lemon, E.R. & Sinclair, T.R. (1978). Leaf elongation and turgor pressure in field-grown soybean. Agron. J., 70, 761-764.
Williams, R.F. (1975). The shoot apex, leaf growth and crop production. J. Aust. Inst. Agric. Sci., 41 18-26.
Wilson, D.R., Fernandez, C.J. & McCree, K.J. (1978). CO_2 exchange of subterranean clover in variable light environments. Crop Sci., 18, 19-22.

Wilson, J.R. & Ludlow, M.M. (1983). Time trends of solute accumulation and the influence of potassium fertiliser on osmotic adjustment of water stressed leaves of three tropical grasses. Aust. J. Plant Physiol., 10, 523-537.

Woodward, F.I. (1983). Instruments for the measurement of photosynthetically active radiation and red, far-red and blue light. J. App. Ecol., 20, 103-116.

Yamazaki, K. (1963). Studies on leaf formation in rice plants. II. The development of leaves in relation to their position on a stem. Proc. Crop Sci. Soc. Japan, 32, 81-88.

Yamazaki, K. (1964). Studies on leaf formation in rice plants. IV. Some experiments revealing the role of processes in leaf formation. Proc. Crop Sci. Soc. Japan, 33, 237-242.

INDEX

SOCIETY FOR EXPERIMENTAL BIOLOGY SEMINAR SERIES

A series of multi-author volumes developed from seminars held by the Society for Experimental Biology. Each volume serves not only as an introductory review of a specific topic, but also introduces the reader to experimental evidence to support the theories and principles discussed, and points the way to new research.

1. Effects of air pollution on plants Edited by T.A. MANSFIELD
2. Effects of pollutants on aquatic organisms Edited by A.P.M. LOCKWOOD
3. Analytical and quantitative methods Edited by J.A. MEEK and H.Y. ELDER
4. Isolation of plant growth substances Edited by J.R. HILLMAN
5. Aspects of animal movement Edited by H.Y. ELDER and E.R. TRUEMAN
6. Neurones without impulses: their significance for vertebrate and invertebrate systems Edited by A. ROBERTS and B.M.H. BUSH
7. Development and specialisation of skeletal muscle Edited by D.F. GOLDSPINK
8. Stomatal physiology Edited by P.G. JARVIS and T.A. MANSFIELD
9. Brain mechanisms of behaviour in lower vertebrates Edited by P.R. LAMING
10. The cell cycle Edited by P.C.L. JOHN
11. Effects of disease on the physiology of the growing plant Edited by P.G. AYRES
12. Biology of the chemotactic response Edited by J.M.LACKIE and P.C. WILLIAMSON
13. Animal migration Edited by D.J. AIDLEY
14. Biological timekeeping Edited by J. BRADY
15. The nucleolus Edited by E.G. JORDAN and C.A. CULLIS
16. Gills Edited by D.F. HOULIHAN, J.C. RANKIN and T.J. SHUTTLEWORTH
17. Cellular acclimatisation to environmental change Edited by A.R. COSSINS and P. SHETERLINE
18. Plant biotechnology Edited by S.H. MANTELL and H. SMITH
19. Storage carbohydrates in vascular plants Edited by D.H. LEWIS
20. The physiology and biochemistry of plant respiration Edited by J.M. PALMER
21. Chloroplast biogenesis Edited by R.J. ELLIS
22. Instrumentation for environmental physiology Edited by B. MARSHALL and F.I. WOODWARD
23. The biosynthesis and metabolism of plant hormones Edited by A. CROZIER and J.R. HILLMAN
24. Coordination of motor behaviour Edited by B.M.H. BUSH and F. CLARAC
25. Cell ageing and cell death Edited by I. DAVIES and D.C. SIGEE
26. The cell division cycle in plants Edited by J.A. BRYANT and D. FRANCIS
27. Control of leaf growth Edited by N.R. BAKER, W.J. DAVIES and C. ONG
28. Biochemistry of plant cell walls Edited by C.T. BRETT and J.R. HILLMAN